# Inventing a European Nation

Engineers for Portugal, from Baroque to Fascism

# Synthesis Lectures on Global Engineering

Editor
**Gary Downey**, *Virginia Tech*

Assistant Editor
**Kacey Beddoes**, *Oregon State University*

The Global Engineering Series challenges students, faculty and administrators, and working engineers to cross the borders of countries, and it follows those who do. Engineers and engineering have grown up within countries. The visions engineers have had of themselves, their knowledge, and their service have varied dramatically over time and across territorial spaces. Engineers now follow diasporas of industrial corporations, NGOs, and other transnational employers across the planet. To what extent do engineers carry their countries with them? What are key sites of encounters among engineers and non-engineers across the borders of countries? What is at stake when engineers encounter others who understand their knowledge, objectives, work, and identities differently? What is engineering now for? What are engineers now for?

The Series invites short manuscripts making visible the experiences of engineers and engineering students and faculty across the borders of countries. Possible topics include engineers in and out of countries, physical mobility and travel, virtual mobility and travel, geo-spatial distributions of work, international education, international work environments, transnational identities and identity issues, transnational organizations, research collaborations, global normativities, and encounters among engineers and non-engineers across country borders.

The Series juxtaposes contributions from distinct disciplinary, analytical, and geographical perspectives to encourage readers to look beyond familiar intellectual and geographical boundaries for insight and guidance. Holding paramount the goal of high-quality scholarship, it offers learning resources to engineering students and faculty and working engineers crossing the borders of countries. Its commitment is to help them improve engineering work through critical self-analysis and listening.

Inventing a European Nation: Engineers for Portugal, from Baroque to Fascism

Maria Paula Diogo and Tiago Saraiva

ISBN: 978-3-031-01001-9    paperback
ISBN: 978-3-031-02129-9    ebook
ISBN: 978-3-031-00173-4    hardcover

DOI 10.1007/978-3-031-02129-9

A Publication in the Springer series
*SYNTHESIS LECTURES ON GLOBAL ENGINEERING*

Lecture #6
Series Editor: Gary Downey, *Virginia Tech*
Series Assistant Editor: Kacey Beddoes, *Oregon State University*
Series ISSN
Print 2160-7664    Electronic 2160-7672

# Inventing a European Nation

## Engineers for Portugal, from Baroque to Fascism

Maria Paula Diogo
NOVA School of Science and Technology

Tiago Saraiva
Drexel University

*SYNTHESIS LECTURES ON GLOBAL ENGINEERING #6*

# ABSTRACT

This book deals with the simultaneous making of Portuguese engineers and the Portuguese nation-state from the mid seventeenth century to the late twentieth century. It argues that the different meanings of being an engineer were directly dependent of projects of nation building and that one cannot understand the history of engineering in Portugal without detailing such projects. Symmetrically, the authors suggest that the very same ability of collectively imagining a nation relied on large measure on engineers and their practices. National culture was not only enacted through poetry, music, and history, but it demanded as well fortresses, railroads, steam engines, and dams.

Portuguese engineers imagined their country in dialogue with Italian, British, French, German or American realities, many times overlapping such references. The book exemplifies how history of engineering makes more salient the transnational dimensions of national history. This is valid beyond the Portuguese case and draws attention to the potential of history of engineering for reshaping national histories and their local specificities into global narratives relevant for readers across different geographies.

# KEYWORDS

Portugal, engineering and nation building, military engineers, civil engineers, railways, dams, colonial engineering

# Contents

CHAPTER 1

# Introduction

## 1.1 WHY TO WRITE AND READ A BOOK ON PORTUGUESE ENGINEERS?

This book deals with the simultaneous making of Portuguese engineers and the Portuguese nation-state from the mid-17th century to the mid-20th century. We argue that the different meanings of being an engineer in this period were directly dependent on projects of nation building and that one cannot understand the history of engineering in Portugal without looking in detail at such projects. Symmetrically, we suggest that the very same ability of collectively imagining a nation relied in large measure on engineers and their practices. National culture was not only enacted through poetry, music, and history, but it also demanded fortresses, railroads, steam engines, and dams. The focus on engineers, on their training and sociability as well as on the technological artifacts they were associated with, promises a renewed understanding of the nature of nation-building processes.

It might seem ironic that the co-authors of this book, two scholars deeply engaged in exploring transnational dynamics in history, ended up writing a text dedicated to the national experience of engineers in a small western European country. After advocating the need for crossing national boundaries to make sense of history, this can be misunderstood as a step back toward more conservative forms of scholarship that have kept the nation as their natural frame of reference. But our point is that in historicizing the process of nation making through engineering, literally of engineering the nation, the very same national frame is put into question and loses its supposed stability.[1] As the reader will see in the following pages, engineers imagined Portugal by making reference namely to Italian, British, French, German, or American realities, many times overlapping such references. In other words, the history of engineering makes more salient the transnational dimensions of Portuguese history. This is also valid beyond the Portuguese case and draws attention to the potential of history of engineering for reshaping national histories and their local specificities into global narratives relevant for readers across different geographies. In what follows, there will be many particular contexts unfamiliar to readers less acquainted with Portuguese history. However, in bringing engineers to the forefront of the narrative we aim at making such contexts more resonating with historical realities common to other areas of the globe.

---

[1] The growing interested for writing global histories of the nation can be exemplified by Patrick Boucheron, Ed., *Histoire Mondiale de la France*, (Paris: Seuil, 2017); Steven Hahn, *A Nation Without Borders: The United States and World in an Age of Civil Wars, 1830–1910*, (New York: Viking, 2016).

In fact, one of the distinguishable characteristics of being a Portuguese engineer was to be part of an international engineering community. International recognition was almost a precondition to speak with authority in the national context, a reality not limited, again, to the Portuguese case. Engineers have a long-lasting tradition of crossing geographical borders and presenting themselves as part of a large transnational community. Although this tradition goes back to networks of artisans, stonemasons, and architects who traveled and exchanged correspondence since the Middle Ages and the Renaissance, it was during the 18th century that the increasing levels of specialization led to the definition of the profile of the modern European engineer as an expert with a specific, unique, and exclusive body of knowledge, based on scientific training directed at the solution of practical problems.

As part of the series "Global Engineering," *Engineers for Portugal* wishes to contribute to a better understanding of the diversity of environments shaping the identity of engineers in different places and time settings, asking why, in spite of local and national differences, there is a sense of common heritage, a sort of transnational "Republic of Engineers" that prevails over borders and centuries.[2] In this sense, this volume will also contribute to a wider understanding of how engineers built a professional tradition by intertwining local, national, and global dimensions.

## 1.2   METHODOLOGICAL QUESTIONS

Much has been written on the history of engineering in Great Britain, France, Germany, or the United States.[3] Comparatively, the bibliography on the so-called countries of the European periphery is still relatively scarce.[4] In the early 21st century research networks such as STEP (Science and Technology in the European Periphery) and ToE (Tensions of Europe) actively

[2] We use the concept of "Republic of Engineers" as an analogy to the concept of Republic of Letters (*Respublica literaria*). The Republic of Letters emerged in the 17th century as a transnational self-proclaimed community of scholars and literary figures. See, among many others, Lorraine Daston, "The ideal and reality of the republic of letters in the enlightenment," *Science in Context 4.2*, (1991), pp. 367–386; Marc Fumaroli, "The republic of letters." *Diogenes 36*, no. 143 (1988), pp. 129–152.

[3] Among a vast literature see, Antoine Picon, *L'invention de l'ingénieur moderne. L'Ecole des Ponts et Chaussées, 1747–1851*, (Paris: Presses de l'ENPC, 1992); Ken Alder, *Engineering the Revolution. Arms and Enlightenment in France, 1763–1815*, (Princeton, NJ: Princeton University Press, 1997); Hélène Vérin, *La Gloire des Ingénieurs. L'intelligence technique du XVe au XVIIIe Siècle*, (Paris, Albin Michel, 199); Bruno Belhoste, *La Formation d'une Technocratie. L'École Polytechnique et ses élèves de la Révolution au Second Empire*, (Paris: Belin, 2003); André Grelon, Ed., *Les Ingénieurs de la Crise. Titre et Profession Entre les Deux Guerres*, (Paris, 1986); Robert A. Buchanan, *The Engineers: A History of the Engineering Profession in Britain, 1750–1914*, (London: Jessica Kingsley Publishers, 1989); Terry S. Reynolds, Ed., *The Engineer in America: A Historical Anthology from Technology and Culture*, (Chicago: Chicago University Press, 1991); Edwin T. Layton, *The Revolt of the Engineers*, (Baltimore, MD: John Hopkins University Press, 1986); David Noble, *America by Design: Science, Technology, and the Rise of Corporate Capitalism*, (Oxford and New York: Oxford University Press, 1979); Eda Kranakis, *Constructing a Bridge. An Exploration of Engineering Culture, Design and Research in 19th-Century France and America*, (Cambridge, MA: The MIT Press, 1997); Kees Gispen, *New Profession, Old Order: Engineers and German Society, 1815–1914*, (Cambridge: Cambridge University press, 1989); Gerd Hortleder, *Das Gesellschaft des Ingenieurs: Zum Politischen Verhalten der Technischen Intelligenz in Deutschland*, (Frankfurt: Suhrkamp, 1970).

[4] The following collections are important exceptions: K. Chatzis, Ed., "The national identities of engineers," *History and Technology 23*, (2007); Robert Fox and Anna Guagnini, Eds., *Education, Technology and Industrial Performance in Europe, 1850–1939*, (Cambridge, Cambridge University Press, 1993); A. C. Matos, M. P. Diogo, I. Gouzevitch, and A. Grelon, Eds.,

contributed to change this status-quo, by presenting case studies on countries that were usually considered mere copies of the canonical cases.[5] The intellectual agenda of such networks produced a powerful critique of diffusion models through stressing the appropriation and circulation of knowledge as creative processes.[6] It aimed at bringing to the forefront the agency of local sites and of local actors in shaping the way expertise was used in different communities in a continuous process of negotiation about appropriation. Historians committed to this agenda proposed a "shift from the point of view of *what has been transmitted* to the view of *how, what was received has been appropriated*."[7] Instead of asking for the lack of technology or engineers in the peripheries, it has proved much more productive to investigate the role of engineering and technology in the history of the peripheries.[8]

It is no coincidence that much of the historiography on Portuguese engineering we build on here was developed by scholars active in the ToE and STEP networks.[9] The granting of historical relevance to processes of appropriation has enabled new historical questions that go beyond the traditional issues of when did a certain theory or machine developed in central coun-

---

*The Quest for a Professional Identity: Engineers Between Training and Action*, (Lisbon: Colibri, 2009); A. Lafuente, A. C. Matos, and T. Saraiva, *Maquinismo Ibérico*, (Aranjuez: Doce Calles, 2007).

[5] STEP (Science and Technology in the European Periphery) emerged as a multi-national research group focused on the study of processes and models of circulation of scientific and technological knowledge "in contexts regarded traditionally as peripheral." STEP was founded in May 1999, in Barcelona, and gathers researchers and university teachers from Belgium, Denmark, Finland, Greece, Hungary, Italy, Portugal, Russia, Spain, Sweden, and Turkey (http://147.156.155.104/). ToE (Tensions of Europe) is "an international scientific network consisting of almost 300 scientists from all over Europe and the United States. Since the start, also in 1999, the network explored a broad range of themes, focusing on the linking and delinking of infrastructures, the emergence of transnational technical communities and the circulation of artifacts, systems, knowledge and people." (http://www.tensionsofeurope.eu/). The research network eventually led to the six volume book series Making Europe, edited by Palgrave Macmillan http://www.makingeurope.eu/www/bookseries.

[6] T. Misa and J. Schot, "Introduction—inventing Europe: Technology and the hidden integration of Europe," *History and Technology 21*, 1 (March 2005), pp. 1–20, 8.

[7] K. Gavroglu, M. Patiniotis, F. Papanelopoulou, A. Simões, A. Carneiro, M. P. Diogo, J. Ramon Bertomeu-Sánchez, A. Garcia Belmar, and A. Nieto-Galan, "Science and technology in the European periphery. Some historiographical reflections," *History of Science 46*, 152 (2008), pp. 153–175.

[8] D. Edgerton, *The Shock of the Old: Technology and Global History since 1900*, (Oxford: Oxford University Press, 2006); "Creole technologies and global histories: Rethinking how things travel in space and time," *HoST 1* (2007), pp. 75–112. Edgerton prefers the terminology of rich and poor countries to that of center and periphery. See also the important essay by Hyungsub Choi, "The social construction of imported technologies: Reflections on the social history of technology in modern Korea," *Technology and Culture 58*, (2017), pp. 905–920.

[9] See namely, M. P. Diogo, A construção de uma identidade profissional: A associação dos engenheiros civis portugueses (1869–1937), Ph.D. thesis (Lisbon, New University of Lisbon, 1994); A. C. Matos, *Ciência, Tecnologia e Desenvolvimento Industrial no Portugal Oitocentista: O Caso dos Lanifícios do Alentejo*, (Lisbon: Estampa, 1998); T. Saraiva, *Ciencia y Ciudad. Madrid y Lisboa, 1851–1900*, (Madrid: Ayuntamento de Madrid, 2005); A. P. Silva and M. P. Diogo (2006), "From host to hostage. Portugal, Britain and the telegraph networks." In: *Networking Europe. Transnational Infrastructures and the Shaping of Europe, 1850–2000*. Ed., by E. van der Vleuten and A. Kaijser. (Sagamore Beach: Science History Publications), pp. 51–69; M. P. Diogo and A. C. Matos, "Aprender a ser ingeniero. La enseñanza de la ingeniería en el Portugal de los siglos XVIII y XIX" in A. Lafuente, A. C. de Matos, and T. Saraiva, Eds., *Maquinismo Ibérico*, (Madrid: Ediciones Doce Calles, 2007), pp. 123–45; Ana Cardoso Matos and Maria Paula Diogo, "Le role des ingénieurs dans l'administration portuguaise: 1852–1900," *Quaderns d'Història de l'Enginyeria*, vol. X, (2009), pp. 351–365; M. Macedo, *Projectar e Construir a Nação. Engenheiros, Ciência e Território em Portugal no Século XIX*, (Lisbon: ICS, 2012); Maria Luísa Sousa, La mobilité automobile au Portugal. La construction du systeme socio-technique, 1920–1950. Ph.D. dissertation Université de la Sorbonne nouvelle—Paris III, 2013; Ana Carneiro, Teresa Salomé Mota, and Vanda Leitão, *O Chão Que Pisamos: A Geologia ao Serviço do Estado, 1848–1974*, (Lisbon: Colibri/CIUCHT, 2013).

tries arrive in a peripheral country like Portugal, or what expertise did Portuguese engineers lack when compared to those from the centers. The increasingly ambitious scholarship of Portuguese historians of technology in the last two decades made possible the big picture offered in this book, the first single volume dealing with the history of Portuguese engineering across three centuries.[10]

*Engineers for Portugal* is conceived at the crossroads of history of engineering, political history, and cultural history. It analyzes Portuguese engineers and engineering in specific historical contexts, as well as the way they forged a strong professional identity simultaneously as efficient technicians and trustworthy decisio makers. Indeed, in many of the examples that we offer in this book, engineers were more than a powerful auxiliary of kings, politicians, or capitalists. They assumed themselves the forefront role as ministers, courtiers, influential public intellectuals, or as entrepreneurs. *Technopolitics* is now a common way of referring in history of technology to "the strategic practice of designing or using technology to enact political goals."[11] Less explored are the processes studied in this book through which engineers contribute to define such political goals, historical actors responsible for enlarging the national political imagination.[12] In a suggestive synthesis, Ken Alder demonstrated how engineers were historically "designed to serve," attaching their existence to issues and problems beyond their own interests.[13] It is thus no surprise to find out that "the identities of engineers have been linked in important ways to nations and nation states."[14] Building on this literature, we don't take the entities engineers served—kings, states, or corporations—as prexistant stable historical realities, preferring instead to explore how engineers' practices contributed to the historical processes of making and sustaining such entities.

[10] Previous valuable attempts have relied heavily on economic history and contributions from engineers themselves. See namely the edited volumes that were produced by occasion of the major exhibition Engenho e Obra held in Lisbon in 2002: J. M. Brandão de Brito, M. Heitor, and M. F. Rollo, Eds., *Engenharia em Portugal no Século XX*, 3 vols. (Lisbon: Dom Quixote, 2004). From a sociology of professions point of view see also, Maria Lurdes Rodrigues, *Os Engenheiros em Portugal*, (Oeiras: Celta, 1999).

[11] The term technopolitics was used by G. Hecht, *The Radiance of France: Nuclear Power and National Identity*, (Cambridge, MA: MIT Press 1998), 15. For a different use of the same concept, Timothy Mitchell, *Rule of Experts: Egypt, Techno-Politics, Modernity*, (Berkeley: University of California Press, 2002). It resonates with the terms techno-diplomacy, first used by J. Der Derian, *On Diplomacy: A Genealogy of Western Estrangement*, (Oxford: Basil Blackwell, 1987), and techno-economics widely used by economists and historians of technology.

[12] This argument is more fully developed in, T. Saraiva, "Inventing the technological nation: The example of Portugal (1851–1898)," *History and Technology 23* (2007), pp. 263–73. Other examples of scholarship dealing explicitly with the relation between nation and technology are, Chandra Mukerji, *Impossible Engineering: Technology and Territoriality on the Canal du Midi*, (Princeton, NJ: Princeton University Press, 2016); Judith Schueler, *Materializing Identity: The Co-Construction of the Gothard Railway and Swiss National Identity*, (Aksant, 2008); Sara Pritchard, *Confluence: The Nature of Technology and the Remaking of the Rhone*, (Cambridge, MA: Harvard University Press, 2011); Johan Schot, Harry Lintsen, and Arie Rip, Eds., *Technology and the Making of the Netherlands: The Age of Contested Modernization, 1890–1970*, (Cambridge, MA: MIT Press, 2010); David E. Nye, *American Technological Sublime*, (Cambridge, MA: MIT Press, 1994); Edward Jones-Imhotep, *The Unreliable Nation: Hostile Nature and Technological Failure in the Cold War*, (Cambridge, MA: MIT Press, 2017).

[13] K. Alder, "French engineers become professionals, or, how meritocracy made knowledge objective," in W. Clark, J. Golinski, and S. Shaffer, Eds., *The Sciences in Enlightened Europe*, (Chicago: University of Chicago Press, 1999), pp. 94–125.

[14] G. L. Downey and J. C. Lucena, "Knowledge and professional identity," *History and Technology 20*, (2004), pp. 393–420.

Continuities between engineering, politics, and culture tend to be more salient in alleged peripheral countries such as Portugal.[15] Well-established institutions, enjoying sustainable funding and long permanence in time, reproducing growing numbers of technically trained experts, are more prone to drive historians into the path of writing the history of engineering as a teleological narrative of the formation of an independent social field with its own rules for attributing social value and practices of social reproduction. The typical multiplicity of roles found in poorer countries in which actors have less defined identities, funds are scarcer, and institutional instability is the rule, demands instead from the historian of engineering an engagement with political and cultural history more pronounced than the one normally found in texts dealing with French, British or American engineers. We suggest that this apparent disadvantage of writing histories of engineering from the periphery might constitute a privileged position to explore the deep entanglements of the simultaneous making of engineers and society.

## 1.3 THE STRUCTURE OF THE BOOK

This volume covers around 350 years of Portuguese history, a period during which Portugal was successively ruled by absolutist (1640–1807), liberal (1820–1910), republican (1910–1926), and authoritarian (1926–1974) governments. In each regime, different emphasis was given to engineers and engineering in such a way that discourses and practices were often shaped by, and concurred to shape, the ideologies and policies of these very regimes. As it is obviously impossible to write a comprehensive history of Portuguese engineering in a short volume, we made a set of choices, focusing on specific topics—embodied by specific engineers, institutions and technologies—that we deem critical, although not exclusive, to understand the identity of Portuguese engineers and their role in defining Portuguese identity.

The justification for starting in 1640 is to underline the coincidence between the beginning of formal training of engineers in Portugal and the recovery of independence of Portugal from the Spanish crown. As we will see in Chapter 2, 17th- and 18th-century seminal textbooks built the profile of Portuguese engineers, separating them not only from men of practice, but also from those engineers ready to be hired by a wealthy patron independently of national allegiance. The first Portuguese engineers were cultivated men able to formalize practical knowledge and put it at the service of a military infrastructure of fortresses in Portugal and across its empire, namely in Brazil, that made plausible the vision of an independent country from the Spanish monarchy. With the arrival in the country of immense riches from Brazilian gold mines since the end of the 17th century, engineers would prove their value in enacting the baroque visions of D. João V for Lisbon—the capital of the country—as a new Rome, beacon of Christianity, progressively adding to their initial military identity the role of courtier.

In Chapter 3, we explore how new institutions—the *Escola Politécnica* (Polytechnic School) and the *Escola so Exército* (Army School)—although built on the previous experience

---

[15] See, for the Portuguese case, Marta Macedo and Jaume Valentines-Álvarez, "Technology and nation: Learning from the periphery," *Technology and Culture* 57, (2016), pp. 989–997.

of engineering training of the 18th century, produced a new kind of engineer with expertise in basic sciences such as mathematics, chemistry, or physics, but also knowledgeable in political economy. Portuguese engineers became the perfect example of Pierre Bourdieu's *noblesse d' état*, as asserted by our detailed exploration of the Associação dos Engenheiros Civis Portugueses (Portuguese Association of Civil Engineers).[16]

Chapter 4 further examines how did Portuguese engineers of the second half of the 19th century become a singularly powerful political and cultural elite unparalleled in the European countries they used as role models. New institutions such as the Lisbon *Instituto Industrial* (Industrial Institute) promised to overcome social antagonisms training a new class of industrialists that would bring together the interests of factory owners and workers. This constituted as well the first major involvement of Portuguese engineers with the industrial world, a major addition to engineering practices previously focused on territorial management. By detailing the role of the engineers who directed the Industrial Institute in defining new forms of urban life in the second half of the 19th century, we make the case for considering engineering a significant element of Lisbon's romantic culture of the second half of the 19th century.

In addition to the national territory and urban centers, Portuguese colonies in Africa became since the 1870s a new arena in which engineers tried to assert their allegedly indispensable presence. In Chapter 5, we discuss the colonial visions of engineers invested in transferring their national policies of "material improvements" namely to Angola and Mozambique. Myriads of plans aimed at transforming African colonies into a new Brazil, supplier of key commodities and importer of manufactured products. Engineers, once more, were key historical actors in putting forward such visions, organizing Public Works expeditions that promised to affirm the country's position among European powers in the context of the colonial partition of Africa of the late 19th century.

As we discuss in Chapter 6, the economic crisis of the 1890s meant an important change in Portuguese engineers priorities directing their efforts to the development of national resources at the expense of the focus on transportation infrastructure. While civil engineering kept its importance, new specialized areas, namely chemical and electrical engineering, would be the focus of reforms that would bring the Industrial Institute to the forefront of engineering education in the country, taking that position from the Army School and the Polytechnic School. In 1911, the Lisbon Industrial Institute was renamed *Instituto Superior Técnico* (Technical Superior Institute), still today one of the most influential engineering schools of the country.

Chapter 7 deals with the implementation of such plans under the dictatorial regime that ruled the country from 1926 until 1974. Contrary to the still pervasive views in Portuguese historiography that tend to stress the traditionalist ideology of the New State dictatorship, we highlight the historical role of engineers in making plausible the radical nationalistic ideology of the Portuguese New State. Keeping with the methodology of all other chapters, we empha-

---

[16] Pierre Bourdieu's notion of noblesse d'état refers to modern elites that assert themselves as centers of power. The term is used to distinguish these new actors from the traditional aristocracy based on family genealogy, the so-called noblesse de robe. P. Bourdieu, *La Noblesse d'État. Grandes écoles et esprit de corps* (Paris: Éditions de Minuit, 1989).

size the historical entanglements between changes in engineering practices and major political changes.

In all chapters we pay close attention to the connections between Portuguese engineers and their international counterparts. Dutch methods of fortification, French polytechnic education and Saint Simonianism, German schools of engineering, and American engineering research laboratories, all play an important role in our narrative. As important as the dialogue between Portuguese engineers and their international counterparts is their role in colonial settings. Adding to the interest of the Portuguese case, engineers not only perceived themselves as being part of the European periphery, but they strived as well to assert Portugal as an imperial entity, first centered in Brazil, and since the second half of the 19th century, in Africa. The reader will thus find out throughout the entire narrative discussions of how important projects in the Portuguese colonies were for the shared sense of identity of Portuguese engineers. After these introductory remarks, it will constitute no surprise for the reader to learn that such discussions will also lead us to highlight the contribution of Portuguese engineers in defining the way the colonies became an integral part of Portuguese national identity.

The conclusion briefly highlights how new institutions (New University of Lisbon, University of Aveiro) and engineering fields such as Information Technology (IT) and environment contributed to the processes of democratization after the 1974 revolution and integration in the European Union since the 1980s. We summarize the findings of previous chapters suggesting how historical knowledge of Portuguese engineering might be inspiring for current and future members of the profession, and calling attention to how throughout the entire history of the country the solving of technological challenges faced by engineers has always been indistinguishable from the solving of political, social, and cultural challenges.

CHAPTER 2

# Making Engineers Portuguese

Although the first official references regarding the use of the title "engineer" in Portuguese language date back to 1559, Portuguese engineers took time to assert their professional skills in a world where circulation of experts was common and competition for a private patron or the support of a foreign court was fierce. As in most other European countries in the 16th, 17th, and 18th centuries, Portuguese engineers were military engineers, whose training was closely connected to their everyday practice in the field. Gradually the importance of a more formal education was acknowledged and engineers would master a wide *corpus* of knowledge, based namely on Italian, French, German, and Dutch treatises.

## 2.1 THE ARTILLERY AND GEOMETRY CLASS (1641) AND THE RESTORATION OF NATIONAL INDEPENDENCE

The beginning of formal training of engineers in Portugal is directly associated with the Restoration of the country's independence from the Spanish crown in 1640 and the war that followed.[1] The simultaneous rebellions of Naples and Catalonia made salient the many tensions of a Spanish Habsburg Empire stretching from Macao to Mexico and from Flanders to Mozambique.[2] But while those separatist movements from Spain failed, the Portuguese cause for independence succeeded in large measure due to the largesse of funds arriving from Brazil's sugar plantations able to sustain an enlarged army.[3] The incapacity of the overextended Spanish empire to protect Portuguese holdings in Brazil from Dutch and French ambitions constituted one of the main grievances expressed by Portuguese elites who had their major source of income in the Brazilian 350 slave plantations producing annually some 20,000 tons of sugar.

---

[1] From 1580, when the Portuguese king D. Sebastião died heirless, until 1640, Portugal lived under the Spanish rule in a political arrangement known as the Iberian Union (two crowns, one king). The Portuguese Restoration War lasted for 28 years (1640–1668) and ended with Spanish defeat. Rafael Valladares, *La Rebelión de Portugal. Guerra, Conflicto y Poderes en la Monarquía Hispánica, 1640–1680,* (Valladolid, 1998); António Manuel Hespanha, "Revoltas e revoluções: a resistência das elites provinciais," *Análise Social,* XXVIII, no. 120 (1993), pp 81–103; Fernando Dores Costa, *A Guerra da Restauração, 1641–1668,* (Lisbon: Horizonte, 2004).

[2] J. H. Elliot, *The Count-Duke of Olivares. The Statesman in an Age of Decline,* (New Haven, CT: Yale University Press, 1986); Geoffrey Parker, *War, Climate Change and Catastrophe in the 17th Century,* (New Haven, CT: Yale University Press, 2013); Jean-Frédéric Schaub, *Portugal e a Monarquia Hispânica, 1580–1640,* (Lisbon: Livros Horizonte, 2001); Jean-Frédéric Schaub, "La crise hispanique de 1640. Le modèle des "revolutions peripheriques" en question (note critique)," *Annales,* V, XLIX:1 (1994), pp. 219–239.

[3] Stuart B. Schwartz, "Prata, açúcar e escravos: de como o império restaurou Portugal," *Tempo 12,* (2008), pp. 201–223.

Figure 2.1: **Luís Serrão Pimentel's signature**. Luís Serrão Pimentel was trained as a mathematician at the Jesuit college of Santo Antão in Lisbon. He became the Kingdom's Chief Cosmographer in 1644 and Chief Engineer in 1671.
Arquivo Nacional Torre do Tombo (ANTT), *Conselho de Guerra, Consultas,* mç. 10A, doc. 210.
© Courtesy of ANTT.

To justify an independent Portugal and promote the cause of D. João IV as the new legitimate king, popular millenarianism was blended with sophisticated scholastic arguments advanced by Jesuit priests about the need of popular consent supporting the monarch.[4] But to enact the vision of an independent nation a major mobilization for the building of a military infrastructure was needed. While a line of fortresses along the eastern and northern borders of the country would defend it from Spanish attacks, a new series of strongholds throughout the Empire would guarantee the economic viability of the nation. Adding to this, the defense of the capital city, Lisbon, demanded a new network of fortresses protecting it from enemy forces arriving by land or sea. It was this infrastructure on which Portuguese independence was built that led to the emergence of the first Portuguese engineers.

In 1641, D. João IV, only one year after assuming the Portuguese crown, established in its court the *Aula de Artilharia e de Esquadria* (Class of Artillery and Geometry), the first antecedent to a more formal engineering academy. To direct it, he chose Luís Serrão Pimentel (1613–1679), a mathematician who had studied at the Jesuit *Colégio de Santo Antão*.[5] This Jesuit college located in Lisbon included an important *Aula da Esfera* (Sphere Class), which had imparted since the end of the 16th-century classes of cosmography, astronomy, seamanship, as well as military architecture and was the main center for mathematical training in the country.[6] The hiring of Serrão Pimentel to the new Class of Artillery and Geometry reflected the mobilization of the learned elite for the building of the new state bureaucracy of the Restoration, as materialized

---

[4] Luis Reis Torgal, "A restauração. Breves reflexões sobre a sua historiografia," *Revista de História das Idéias*, I, (1997), pp. 23–40.

[5] For more details on Luís Serrão Pimentel's work on fortification see: Antónia Fialho Conde, "The art of war: Tradition and innovation in the iconographic representation of Alentejo fortresses (17th–18th centuries)," *History Research 3*, 5 (May 2013), pp. 353–364; Nuno Alexandre Martins Ferreira, Luís Serrão Pimentel (1613–1679), Cosmógrafo Mor e engenheiro mor de Portugal (Master dissertation, Faculdade de Letras de Lisboa, 2009); Christovam Ayres de Magalhães Sepulveda, *Historia Organica e Politica do Exercito Português—Provas*, (Lisboa, 1912), vol. 7, p. 61.

[6] Henrique Leitão, "Jesuit mathematical practice in Portugal, 1540–1759," in M. Feingold, Ed., *The New Science and Jesuit Science: 17th Century Perspectives*, (Dordrecht: Kluwer Academic, 2003), pp. 229–247.

namely in the new Overseas Council and War Council responsible for maintaining a more regular and disciplined military.[7] The new Class, located at the royal palace, should guarantee that the military architecture skills previously thought at the Jesuit school were put in direct service of the effort of making a new independent state and establishing a new dynasty.

Six years after, in 1647, Serrão Pimentel earned the support of the king to replace that first Class by a new *Aula de Fortificação e Architectura Militar* (Class of Fortification and Military Architecture), later known as *Academia Militar* (Military Academy). This would become the main institution for training Portuguese engineers until the 19th century. Students at the Class of Military Fortification and Architecture received both theoretical and practical training, the first based on mathematics and geometry and the latter on fieldwork. After concluding their training, students were the only officers holding a military engineering degree forming the core of a new army corps specifically dedicated to engineering.

Although it is not possible to be absolutely sure about the topics taught in this Class, it is plausible to assume that they closely followed Serrão's own book *Methodo Lusitano de desenhar as fortificações das praças regulares e irregulares* (*Portuguese Method for drawing Fortifications*) that was a set text for the students.[8] The *Portuguese Method* was an updated treatise on the art of fortification that draw most of its information from authors such as Freitag, Dogen, Marolois, Goldman, Coheornn, and Stevin who epitomized the so-called Dutch method of fortification. Spanish authors, as well as the new trends of the French school of fortification (Pagan, Bar-le-Duc, Antoine de Ville e Manesson Mallet) were also used.[9] A written treaty clearly differentiated Serrão, the engineer, from other allegedly unlearned men whose practical knowledge could also be used in a battlefield. While the treaty asserted the importance of being knowledgeable about crafts, it denied craftsmen the status of engineer, reserved only to those able to codify such knowledge through written language and mathmatics.[10] In addition to its educational purposes, one should thus keep in mind the work Serrão's book perfomed for the social status of engineer.

Serrão had become familiarized with all those authors through his contact with the military engineers hired during the conflict with Spain by the king D. João IV across Europe, namely Joannes Cieremanns. Trained in mathematics at the Jesuit University of Leuven, Cieremanns would join the Jesuit college of Santo Antão in Lisbon in 1641 where Serrão was a student. Impressed by his military architecture skills, the king would offer Cieremanns a high rank in the Portuguese army and trust him the planning of many of the new fortresses of the border with Spain. Significantly enough, Cieremanns would defect to the Spanish side in 1647. Such acts

---

[7] Luís Reis Torgal, "Restauração e razão de Estado," *Penélope 9/10*, (1993), pp. 163–67; Fernando Dores Costa, "O conselho de guerra como lugar de poder: A delimitação da sua autoridade," *Análise Social*, XLIV no. 191 (2009), pp. 379–414.

[8] Luís Serrão Pimentel, *Methodo Lusitano de Desenhar as Fortificações das Praças Regulares e Irregulares*, (Lisboa, 1680).

[9] Helene Vérin, *La Gloire des Ingénieurs*, L'intelligence technique du XVIe au XVIIIe siècle, (Paris: Albin Michel, 1993), pp. 243–333.

[10] On these interactions between formal and informal knowledge see, Pamela O. Long, *Openess, Secrecy, Authorship: Technical Arts and the Culture of Knowledge from Antiquity to the Renaissance*, (Baltimore, MD: Johns Hopkins University Press, 2001).

Figure 2.2: **Frontispiece of Luís Serrão Pimentel's main book,** *Methodo Lusitanico de desenhar as fortificaçoens ....* **(1680)**.
Combining Dutch, French, and Spanish methods of fortification, Serrão wrote a Portuguese method to train Portuguese engineers. His main challenge was to balance nationalistic aims and international engineering skills in a time where circulation (including defection) of engineers among European courts was common.
© Public domain.

of treason were in fact common among the foreign military engineers hired by the Portuguese king, as also demonstrated by Jean Gillot, another of Serrão's masters, and who had been no less than the Kingdom's Chief Engineer (*Engenheiro-mor*) before defecting to the Spanish cause. Serrão Pimentel would be promoted to this rank of chief engineer in 1663.

It is important to retain these tensions between nationalistic aims and international engineering skills in the institutionalization of engineering training in Portugal. Serrão produced a Portuguese method to train Portuguese engineers which was a combination of Dutch and French methods and that also included methods from Spain, the enemy against which Portuguese engineers were being mobilized. The challenge faced by Serrão Pimentel was how to make Portuguese engineers out of this international body of knowledge that he learned from an international community ready to sell their services to the highest bidder. The Class of Military Fortification and Architecture was part of the answer by making mathematical skills more amenable to the purposes of military defense of the state than the previous lectures at the Jesuit college of Santo Antão. Under direct control of the War Council it was meant to produce engineers identified with the Portuguese cause. In other words, the Restored Portuguese nation state demanded more than engineers, it demanded *Portuguese* engineers. The set of traits first defined at this Class formed a coherent identity of Portuguese engineers that would endure in the next centuries: international body of scientific knowledge; strong emphasis on mathematics; field experience; servers of the state; nationalistic ideology; military ethos.

The task of these first Portuguese military engineers was clear: the planning and building of fortifications enabling Portuguese independence in the metropole and in its empire. At the turn of the century new local Classes, modeled on the Lisbon one, were founded in order to guarantee a stable body of skilled experts serving this military infrastructure. In Portugal, these Classes were located in Viana do Castelo (1701), Elvas (1732), and Almeida (1732), all border cities with newly built defenses planned according to the rules of Serrão's *Portuguese Method of Drawing Fortifications* to defend the country from Spanish invasions. In the Empire, these included cities like Luanda (1699) in Angola, retaken from the Dutch in 1648 and the main port for shipping African slaves to the Americas, namely Brazil. But it was in Brazil itself that the presence of the new Portuguese military engineers was more obvious. New classes were started in Bahia (1696), Rio de Janeiro (1698), São Luis do Maranhão (1699), Recife (1701), and Belém (1758). Some 140 Portuguese military engineers served in Brazil from the 1670s until the 1770s.[11]

---

[11] Beatriz Bueno, "Desenho e desígnio: o Brasil dos engenheiros militares" (Ph.D. dissertation, University of São Paulo, 2001); Dulcyene Maria Ribeiro, "A formação dos engenheiros militares: Azevedo Fortes, matemática e ensino da engenharia militar no século XVIII em Portugal e no Brasil" (Ph.D. dissertation, University of São Paulo, 2009).

## 2.2   MANUEL DE AZEVEDO FORTES AND THE PORTUGUESE ENGINEER

In Lisbon, the Class of Military Fortification and Architecture kept the pace with European developments. From 1686–1703, Manuel de Azevedo Fortes, who later became Chief Engineer of the kingdom (he succeeded Serrão Pimentel in that post), taught mathematics at the military academy. Drawing on his extensive international experience, Azevedo Fortes sketched a first plan for regulating the work of engineers, namely their professional duties and academic training. He is thus a central figure for our understanding of the formation of Portuguese engineers' identity.

Azevedo Fortes (1660–1749) was the illegitimate son of a French nobleman serving at the Court in Lisbon as commander of the French troops (allies against Spain), and a well-born Portuguese lady whose name is not known.[12] Although his parents did not officially recognize Azevedo Fortes as their legitimate son, his father, in the role of a benefactor, took full responsibility for his education. When Azevedo Fortes was only 10 years old, he was taken from the "indulgence, idleness and amusement of the court"[13] and sent first to the Imperial College of Madrid, the center of Jesuit education of the Spanish monarchy, and then to the University of Alcalá de Henares for "instruction in the strict sciences," specifically in the area of philosophy.[14] His academic success led his father to send him to the Plessis College in France to widen his knowledge of modern philosophy, experimental philosophy and mathematics, "not being easy to determine in which of these subjects he would become most eminent."[15] It deserves notice that European rivalries didn't hinder a transnational education in elite institutions across the continent.

Having completed his education, Fortes taught mathematics at the University of Sienna, serving Francesco Maria de Medici, brother of Cosimo de Medici, Grand Prince of Tuscany. Fortes' reputation preceded him when he returned to Portugal, being welcomed enthusiastically into the country's intellectual circles, particularly the one hosted by the Count of Ericeira.[16] The Count of Ericeira's circle was one of the havens where the new European scientific and

---

[12] In official declarations Azevedo Fortes gave several different accounts of his parenthood. However, Joaquim Pereira da Silva Real, who wrote a funeral elegy for Azevedo Fortes that was presented to the *Academia Real de História*, (Royal Academy of History), after listing the different possibilities, concluded that, on the basis of conversations with Fortes himself, he was convinced that Monsieur Lembrancour (later referred to as Monsieur Leblancour) was his father. See Jozé Gomes da Cruz, *Elogio Funebre de Manoel de Azevedo Fortes*, (Lisbon), 1754, pp. 2 and 4. For more details on Manuel de Azevedo Fortes see Maria Paula Diogo, Ana Carneiro, and Ana Simões, "El grand tour de la tecnología: El estrangeirado manuel de Azevedo Fortes," in *Maquinismo Ibérico*, Ed., A. Lafuente, A. Cardoso de Matos, and T. Saraiva (Aranjuez: Doce Calles, 2007), pp. 101–121.

[13] Real, funeral elegy, n/p.

[14] Diogo Barbosa Machado, *Biblioteca Lusitana*, (Lisboa, 1752), vol. 3, p. 186.

[15] Machado, Biblioteca, p. 186.

[16] For more on the Count of Ericeira's circle see Ana Carneiro, Ana Simões, and Maria Paula Diogo,"Enlightenment science in Portugal: the Estrangeirados and their communication networks," *Social Studies of Science 30*, 4 (2000), pp. 591–619; Maria Paula Diogo, Ana Carneiro, and Ana Simões, "Ciência portuguesa no iluminismo. Os estrangeirados e as comunidades científicas europeias," in *Enteados de Galileu? A Semiperiferia no Sistema Mundial de Ciência*, Ed., J. A. Nunes and M. E. Gonçalves (Porto: Edições Afrontamento, 2001), pp. 209–238; Ana Simões, Ana Carneiro, and Maria Paula Diogo, "Constructing knowledge: 18th-century Portugal and the new sciences," *Archimedes*, 2 (1999), pp. 163–197; Ana Simões, Ana

Figure 2.3: **Portrait of Manuel de Azevedo Fortes by the French painter Pierre-Antoine Quillard, c. 1727**.
Manuel de Azevedo Fortes was trained in Mathematics and Philosophy in France and Spain and was part of a Portuguese elite known as *estrangeirados* (European-oriented intellectuals). As the Kingdom's Chief Engineer (1719), Fortes was close to the royal circle and championed engineering as both a technical and a political profession in the service of the absolutist monarch. His portrait, painted by Quillard, the court's favorite painter, underlines the identity of the engineer as courtier.
© Wikimedia Commons. Collection of the Museu de Arte Antiga.

philosophical ideas took root in the country. Fortes soon became one of the count's *protégés* and got the interest of King Pedro II, who offered him a teaching position at the Class of Military Fortification and Architecture, and the post of infantry captain specializing in engineering in 1698 (he was promoted to chief engineering sergeant in 1703).

During his military career, the fact that he spoke fluent Portuguese, Spanish, Italian, and French led him to undertake special assignments in addition to his regular duties relating to fortifications. These included military espionage in Spain and working as a liaison officer and interpreter in dealings between privates and non-commissioned officers from Portugal and commissioned officers from foreign armies.[17] In 1710, he was appointed military governor of the border town of Castelo de Vide, with the rank of colonel, and instructed to rebuild its fortifications. He was made Chief Engineer in 1719[18] and promoted to chief battle sergeant in 1735.[19]

When the Royal Academy of History was created in 1720, Fortes, already Chief Engineer and a leading member of the Count of Ericeira's circle, was one of its founding members. The academy embodied an enlightenment view of historical knowledge and it was charged with writing a new history of Portugal, an important tool for legitimizing the Portuguese monarchy and its imperial possessions beyond previous considerations of divine providence. Fortes was responsible for "questions of the geographical history, both ecclesiastical and secular, of these kingdoms."[20] This led him to write in 1722 the *Tratado do modo o mais facil e o mais exacto de fazer as cartas geographicas*[21] ("Treaty on the simplest and most accurate method of making geographical maps"), an important text in the teaching of engineers, the principal aim of which was "the honor and esteem of this nation, deeming it unbecoming that we should beg from nations ill versed in a science and an art that they themselves learned from the Portuguese nation."[22] The treatise is based on rules set out by Deschales and de Ozanam, and includes complete transcriptions of some passages (acknowledged by Manuel de Azevedo Fortes) from their works.

This *Tratado* takes on a significance that goes beyond its strictly technical component, as it highlights the critical role of engineers as envisioned by Fortes in mapping and managing the kingdom's territory. More so when the Treaty of Utrecht (1713) had replaced the previous pope's prerogative considering territorial disputes between European states and their respective imperial possessions, imposing instead the effective possession of the territory as main criteria

---

Carneiro, and Maria Paula Diogo, "Introductory remarks," in *Travels of learning: A Geography of Science in Europe*, Ed., Ana Simões, Ana Carneiro, and Maria Paula Diogo (Dordrecht: Kluwer Academic Publishers, 2003), pp. 1–18.

[17] Azevedo Fortes infiltrated the Spanish troops disguised as a "Castilian peasant." Cruz, *Elogio Funebre*, 6.

[18] Decree September 23, 1719. Cf. Machado, *Biblioteca*, 188. Sepulveda, *Historia Organica*, 107, indicates 1718 as the date of the promotion.

[19] *Catálogo dos Decretos do Extinto Conselho de Guerra*, 1735, Military History Archives, Batch 94–29 March, no. 6.

[20] Manuel de Azevedo Fortes, *Oração Academica que pronunciou Manoel de Azevedo Fortes na Presença de Suas Magestades hindo a Academia ao Paço, em 22 de Outubro de 1739*, (Lisboa, 1739), 1.

[21] Manuel de Azevedo Fortes, *Tratado do modo o mais facil e o mais exacto de fazer as cartas geographicas, assim da terra como do mar, e tirar as plantas das praças, cidades e edificios com instrumentos, e sem instrumentos para servir de instruccam à fabrica das Cartas Geograficas da Historia Ecclesiastica, e Secular de Portugal*, (Lisboa, 1722).

[22] Manuel de Azevedo Fortes, *Tratado do modo*, 2.

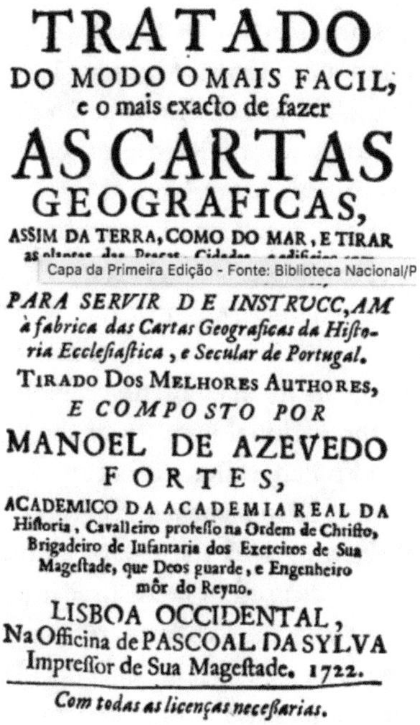

TRATADO
DO MODO O MAIS FACIL,
e o mais exacto de fazer

AS CARTAS
GEOGRAFICAS,

ASSIM DA TERRA, COMO DO MAR, E TIRAR
as plantas das Praças, Cidades, e edificios com

Capa da Primeira Edição - Fonte: Biblioteca Nacional/P

PARA SERVIR DE INSTRUCC, AM
à fabrica das Cartas Geograficas da Historia Ecclefiaftica, e Secular de Portugal.

TIRADO DOS MELHORES AUTHORES,
E COMPOSTO POR

MANOEL DE AZEVEDO
FORTES,

ACADEMICO DA ACADEMIA REAL DA
Hiftoria, Cavalleiro profeffo na Ordem de Chrifto,
Brigadeiro de Infantaria dos Exercitos de Sua
Mageftade, que Deos guarde, e Engenheiro
môr do Reyno.

LISBOA OCCIDENTAL,
Na Officina de PASCOAL DA SYLVA
Impreffor de Sua Mageftade. 1722.

Com todas as licenças neceffarias.

Figure 2.4: **Frontpage of Manuel de Azevedo Fortes'** *Tratado do modo o mais facil e o mais exacto de fazer as cartas geográficas ...* **(1722).**
This work was foundational for the training of Portuguese engineers and was considered essential reading during the whole eighteenth century. Its significance goes beyond its strictly technical component, as it highlights as well the critical role of engineers in mapping and managing the kingdom's territory. The explicit reference to Portugal in the tittle, asserts the national identity of the profession and its strong relationship with the monarchy.
© Public domain.

for the demarcation of borders.[23] Azevedo Fortes work on geographical history at the Academy set the basis for Portuguese control over large tracts of land in the Amazon basin (today Brazil) in dispute with the Spanish vice-royalties of Lima and Santa Fe (respectively, Peru and Colombia). The mobilization of military engineering skills by the Portuguese crown in the area, unmatched by an equivalent effort from the Spanish side, materialized a modern way of making empire.[24] Portuguese engineers surveyed the territory according to the rules taught by Fortes in the *Tratado*, while also putting in place a network of military fortifications confirming Portuguese presence. The new villages and fortresses of Macapá (1752), Chaves (1758), Santarém (1754), Monte Alegre (1758), Alenquer (1758), Óbidos (1758), and Faro (1758) guaranteed the exclusive Portuguese control of 2,400 km of the Amazon river.[25] If fortresses like the one of Macapá never saw military action, they were nevertheless crucial in promoting settlement and organizing the economic activities of their hinterland.[26]

Building on the *Tratado* and as a result of his teaching position at the Class of Military Fortification and Architecture, Fortes wrote *O Engenheiro Português* (*The Portuguese Engineer*) (1728–1729),[27] a seminal work on Portuguese engineering, which defined the profession for the first time in terms of its scientific foundation, clearly separating it from architecture and fully acknowledging social utility as one of its main elements.[28] Note of course, the explicit reference to Portugal in the tittle, asserting the national identity of the profession, a more salient feature considering that Fortes was son of a French military officer and got most of his education abroad.

From a technical point of view, *The Portuguese Engineer* introduced in Portugal the European conventions in technical drawing and the use of standardized codes in treaties, both addressing the instruments and methods used in field surveys and the instruments, conventions and codes used in drawings. Fortes' main sources were, once again, Jacques Ozamam's *Methode de lever les plans et les cartes de terre et de mer,* Dechalles' and Ozamam's *Les Elemens d'Euclide,"* Nicolas Buchotte's *Les Règles dessein et du lavis*, and Jean-Louis Naudin's *L'Ingénieur Français*.

*The Portuguese Engineer* was a didactic work to be used by engineering students[29] and aimed at systematically presenting the essential skills for a Portuguese engineer. The book has two volumes: the first, the *Treatise on Practical Geometry*, is divided into three sections—"the

---

[23] Iris Kantor, "Soberania e territorialidade colonial: Academia Real de História Portuguesa e a América Portuguesa (1720)," *Temas Setecentistas Governo e Populações no Império Portugues*, Ed., Andréa Doré, Antonio Cesar de Almeida Santos (Curitiba Brasil), UFPR/SCHLA-Fundação Araucária, (2009), pp. 233–239.

[24] On the engineering advantage of the Portuguese over the Spaniards see Beatriz Bueno, "A arquitetura das fronteiras do Brasil. Duas faces de um mesmo problema," *Arquitextos*, 13, 148.05, September 2012.

[25] Renata Araújo, "As cidades da Amazônia no século XVIII. Belém, Macapá e Mazagão" (Ph.D. dissertation, Faculdade de Arquitectura, Universidade do Porto, 1998).

[26] Bueno, "A arquitetura."

[27] Manoel de Azevedo Fortes, *O Engenheiro Português: Dividido em Dous Tratados*, (Lisboa, 1728–1729).

[28] Manuel de Azevedo Fortes was one of the key names associated with the institutionalization of engineering in Portugal. See Maria Paula Diogo, A construção de uma identidade profissional—A associação dos engenheiros civis portugueses (Ph.D. dissertation, Faculdade de Ciências e Tecnologia da Universidade NOVA de Lisboa, 1984); "In search of a professional identity: The associacão dos engenheiros civis Portugueses (1869–1937)," *ICON 2*, (1996), pp. 123–137.

[29] The *O Engenheiro Portugues* is defended against critics by the students who used it in *Evidencia Apologetica e Critica Sobre o Primeiro e Segundo Volume das Memorias Militares, pelos Particantes da Academia Militar d' esta Corte*, (Lisboa, 1733).

Figure 2.5: **Frontpage of Manuel de Azevedo Fortes' *O Engenheiro Portuguez* (1728-29), including his portrait.**
The two-volume treaty, mostly based on the French military engineer Marquis de Vauban's doctrine, defined the scientific nature of the engineering profession, demarcating it from architecture and acknowledging its social utility.Note the explicit reference to Portugal in the tittle, asserting the national identity of the new profession.
© Wikimedia Commons.

first on Longimetry, or the measurement of distances: the second on Planimetry, or the measurement of areas: the third on Stereometry, or the measurement of solid bodies;"[30] the second, the *Treatise on Fortification*, is divided into eight sections, the first dealing with general questions relating to the art of fortification, the second with the methods of "the three most celebrated Authors, the Cavalier Antonio de Ville, the Count of Pagan, and the Marshall of France, Monsieur de Vauban, and the 'Fortifications' of an anonymous author who used the former three as his guides,"[31] and the remaining six with specific aspects of fortification.

The aim of the work was clear: to make available in Portuguese language for the country's future engineers the knowledge the Chief Engineer had accumulated internationally during his academic and professional career. Fortes insisted in his unique familiarity with the latest approaches to the subject abroad, which in the training available in Portugal were either non-

[30] Fortes, *Representação*.
[31] Fortes, *Representação*.

TABOADA SEGVNDA.
De partes inteiras, seus primos, & segundos, ou centessimos de parte para a fabrica da Fitta gradual.

| G. | M. | Int. | | G. | M. | Int. | | G. | M. | Int. | | G. | M. | Int. | |
|----|----|----|----|----|----|----|----|----|----|----|----|----|----|----|----|
| 5 | 00 | 43 | 61 | 24 | 00 | 207 | 91 | 43 | 00 | 366 | 50 | 62 | 00 | 515 | 03 |
| 5 | 30 | 47 | 97 | 24 | 30 | 212 | 17 | 43 | 30 | 370 | 55 | 62 | 30 | 518 | 77 |
| 6 | 00 | 52 | 33 | 25 | 00 | 216 | 43 | 44 | 00 | 374 | 60 | 63 | 00 | 522 | 49 |
| 6 | 30 | 56 | 69 | 25 | 30 | 220 | 69 | 44 | 30 | 378 | 64 | 63 | 30 | 526 | 21 |
| 7 | 00 | 61 | 04 | 26 | 00 | 224 | 95 | 45 | 00 | 382 | 68 | 64 | 00 | 529 | 91 |
| 7 | 30 | 65 | 40 | 26 | 30 | 229 | 20 | 45 | 30 | 386 | 71 | 64 | 30 | 533 | 61 |
| 8 | 00 | 69 | 75 | 27 | 00 | 233 | 44 | 46 | 00 | 390 | 73 | 65 | 00 | 537 | 29 |
| 8 | 30 | 74 | 10 | 27 | 30 | 237 | 68 | 46 | 30 | 394 | 74 | 65 | 30 | 540 | 97 |
| 9 | 00 | 78 | 45 | 28 | 00 | 241 | 92 | 47 | 00 | 398 | 74 | 66 | 00 | 544 | 63 |
| 9 | 30 | 82 | 80 | 28 | 30 | 246 | 15 | 47 | 30 | 402 | 74 | 66 | 30 | 548 | 29 |
| 10 | 00 | 87 | 15 | 29 | 00 | 250 | 38 | 48 | 00 | 406 | 73 | 67 | 00 | 551 | 93 |
| 10 | 30 | 91 | 50 | 29 | 30 | 254 | 60 | 48 | 30 | 410 | 71 | 67 | 30 | 555 | 57 |
| 11 | 00 | 95 | 84 | 30 | 00 | 258 | 81 | 49 | 00 | 414 | 69 | 68 | 00 | 559 | 19 |
| 11 | 30 | 100 | 18 | 30 | 30 | 263 | 03 | 49 | 30 | 418 | 65 | 68 | 30 | 562 | 80 |
| 12 | 00 | 104 | 52 | 31 | 00 | 267 | 23 | 50 | 00 | 422 | 61 | 69 | 00 | 566 | 40 |
| 12 | 30 | 108 | 86 | 31 | 30 | 271 | 44 | 50 | 30 | 426 | 56 | 69 | 30 | 569 | 99 |
| 13 | 00 | 113 | 20 | 32 | 00 | 275 | 63 | 51 | 00 | 430 | 51 | 70 | 00 | 573 | 57 |
| 13 | 30 | 117 | 53 | 32 | 30 | 279 | 82 | 51 | 30 | 434 | 44 | 70 | 30 | 577 | 14 |
| 14 | 00 | 121 | 86 | 33 | 00 | 284 | 00 | 52 | 00 | 438 | 37 | 71 | 00 | 580 | 70 |
| 14 | 30 | 126 | 19 | 33 | 30 | 288 | 19 | 52 | 30 | 442 | 28 | 71 | 30 | 584 | 24 |
| 15 | 00 | 130 | 52 | 34 | 00 | 292 | 37 | 53 | 00 | 446 | 19 | 72 | 00 | 587 | 78 |
| 15 | 30 | 134 | 85 | 34 | 30 | 296 | 54 | 53 | 30 | 450 | 09 | 72 | 30 | 591 | 30 |
| 16 | 00 | 139 | 17 | 35 | 00 | 300 | 70 | 54 | 00 | 453 | 99 | 73 | 00 | 594 | 82 |
| 16 | 30 | 143 | 49 | 35 | 30 | 304 | 86 | 54 | 30 | 457 | 87 | 73 | 30 | 598 | 32 |
| 17 | 00 | 147 | 80 | 36 | 00 | 309 | 01 | 55 | 00 | 461 | 74 | 74 | 00 | 601 | 81 |
| 17 | 30 | 152 | 12 | 36 | 30 | 313 | 16 | 55 | 30 | 465 | 61 | 74 | 30 | 605 | 29 |
| 18 | 00 | 156 | 43 | 37 | 00 | 317 | 30 | 56 | 00 | 469 | 47 | 75 | 00 | 608 | 76 |
| 18 | 30 | 160 | 74 | 37 | 30 | 321 | 43 | 56 | 30 | 473 | 31 | 75 | 30 | 612 | 21 |
| 19 | 00 | 165 | 04 | 38 | 00 | 325 | 56 | 57 | 00 | 477 | 15 | 76 | 00 | 615 | 66 |
| 19 | 30 | 169 | 34 | 38 | 30 | 329 | 69 | 57 | 30 | 480 | 98 | 76 | 30 | 619 | 09 |
| 20 | 00 | 173 | 64 | 39 | 00 | 333 | 80 | 58 | 00 | 484 | 80 | 77 | 00 | 622 | 51 |
| 20 | 30 | 177 | 94 | 39 | 30 | 337 | 91 | 58 | 30 | 488 | 62 | 77 | 30 | 625 | 92 |
| 21 | 00 | 182 | 23 | 40 | 00 | 342 | 02 | 59 | 00 | 492 | 42 | 78 | 00 | 629 | 32 |
| 21 | 30 | 186 | 52 | 40 | 30 | 346 | 11 | 59 | 30 | 496 | 21 | 78 | 30 | 632 | 70 |
| 22 | 00 | 190 | 80 | 41 | 00 | 350 | 20 | 60 | 00 | 500 | 00 | 79 | 00 | 636 | 07 |
| 22 | 30 | 195 | 09 | 41 | 30 | 354 | 29 | 60 | 30 | 503 | 77 | 79 | 30 | 639 | 43 |
| 23 | 00 | 199 | 36 | 42 | 00 | 358 | 36 | 61 | 00 | 507 | 52 | 80 | 00 | 642 | 78 |
| 23 | 30 | 203 | 64 | 42 | 30 | 362 | 43 | 61 | 30 | 511 | 29 | 80 | 30 | 646 | 12 |

Figure 2.6: *O Engenheiro Portuguez* (1728-29): **mathematical table**.
For Azevedo Fortes the proficient use of mathematical tables was a practice that defined the professional identity of Portuguese engineers.
© Wikimedia Commons.

existent or extremely out of date.[32] Again, it is important to retain that what qualified Fortes to be in the position of deciding what skills were important to train a Portuguese engineer was his experience outside Portugal.

The didactic character of the work required a special effort by the author to explain the different theories it presented with the utmost clarity, as the book was not meant to be a historical compilation of existing methods, but being instead an account written for practical purposes of "the current doctrine, the most modern, practiced today."[33] The same concern underlies Fortes' obsession with language: he insisted in the importance of precisely defining terms and concepts, considering their mastery essential for the advancement of its practitioners. Finally, both volumes included appendixes and extensive plates illustrating the contents.[34]

Manuel de Azevedo Fortes presented engineering as a multifaceted field of activity ranging from the practical concerns of war (such as how to move an army and solve the logistical problems that arise from its movement) to specifically engineering issues (such as the building of fortifications and the problems relating to material, terrain etc., that this involves) and included as well the equally important task of advising military commanders who had to make decisions on questions of defense. Accordingly, the training of the new engineer combined three areas: scientific training based on mathematics and geometry;[35] practical training on the ground in which theoretical knowledge was tested by carrying out real projects; equipping the engineer to play an active role in decision making.

## 2.3   ENGINEER AND COURTIER

The above reference to Fortes' participation as founding member of the Academy of History created in 1720 also indicates an extended role for engineers in Portuguese society beyond their identification with the defense infrastructure issues of the Restoration years. The Academy constituted a major effort for granting glory to the monarchy, now under the rule of D. João V, epitome of the absolutist King in Portuguese history.[36] Fifty members of the Portuguese intel-

---

[32] When Azevedo Fortes writes "we have little or nothing," he explicitly intends to exempt from his criticism the work by Luís Serrão Pimentel, *Methodo Lusitano de Desenhar as Fortificações das Praças Regulares e Irregulare*, published posthumously (1680).

[33] Fortes, *O Engenheiro Portugues*, "Prologue to the Reader."

[34] For Azevedo Fortes the illustrations played a critical role in ensuring that the theories being expounded were better and more thoroughly understood. On the importance of drawing in science and engineering, see M. F. Faria, *A Imagem Útil: José Joaquim Freire (1760–1847) Desenhador Topográfico e de História Natural: Arte, Ciência e Razão de Estado no Final do Antigo Regime*, (Lisboa: UAL, 2001).

[35] For Azevedo Fortes these were the foundations of science. In *Oração Academica*, he states that "There is certainly no natural science so clear, so exact and thus so perfect as the science of numbers of geometry," (p. 8). Mathematics and geometry are also of decisive importance in the concept of practical science: to quote again from *Oração Academica*, "The great multitude of excellent inventions to be found in the science that diminish the labor of men, allowing them to savour more enjoyment from a more restful life, are enough for it to earn the respect and application it deserves; because without the help of mathematics it would be impossible for artisans to contrive so many and such varied devices by which with little effort solid bodies of prodigious volume and excessive weight can be raised, lowered and suspended with perfect balance," (p. 10).

[36] Isabel Ferreira da Mota, *A Academia Real da História. Os Intelectuais, o Poder Cultural e o Poder Monárquico no Século XVIII*, (Coimbra: Ed. Minerva, 2003).

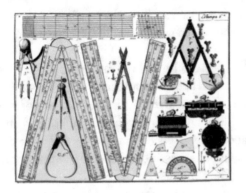

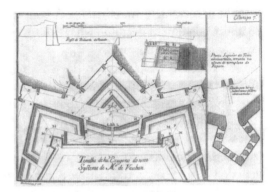

Figure 2.7: *O Engenheiro Portuguez* (1728-29): instruments.
Azevedo Fortes' book introduced Portuguese engineering students to European conventions in technical drawing and the use of standardized codes in treaties, addressing the instruments and methods used in field surveys. © Wikimedia Commons.

Figure 2.8: *O Engenheiro Portuguez* (1728-29): fortification plans. This image illustrates Vauban's fortifications and their importance for the training of Portuguese engineers. Based largely on the work of Antoine de Ville and Blaise Pagan (also used by Fortes in *O Engenheiro Portuguez*), Vauban used the famous 'star-shape' bastion fort design. Two subsequent adaptations, or "systems", strengthened the defensive walls and shielded the interior of the fortification. © Wikimedia Commons.

lectual elite were assembled to write the history of the country in all its dimensions (political, military, ecclesiastical, natural….) asserting the legitimacy of the nation and its unique position in the world.[37] It is important to retain this continuum of activities by Fortes, with no clear differentiation between history and engineering, developing methods for drawing geographical maps as part of the task of writing the history of the monarchy to better serve D. João V, the Magnanimous. Engineering and history were indeed part of a continuum of practices for establishing the global role of the Portuguese absolutist monarchy.

The discovery of copious gold ores in the interior of Brazil, which started to arrive in Portugal in large quantities by the last decade of the 17th century, made plausible the unbridled visions of a king who dreamed of transforming the Portuguese Empire in no less than a new Roman Empire.[38] Engineers would seize the opportunity to assert the status of their skills by demonstrating their value in materializing such reveries, not only through their contribution for

---

[37] Norberto Ferreira da Cunha, "A desdivinização do mundo histórico no século XVIII." A Academia Portuguesa de História (1720–1737), *Diacrítica*, 6 (1991), pp. 249–290.

[38] Stuart B. Schwartz, "De ouro a algodão: A economia brasileira no século XVIII," in *História da Expansão Portuguesa*, Ed., Francisco Bethencourt and Kirti Chauduri (Lisbon: Círculo de Leitores, 1998); Angela Delaforce, "Lisbon, 'This new

the expansion and defense of the empire through maps and fortresses, but also by contributing to the grand plans of transforming Lisbon into a new Rome.

In 1716, a Portuguese lavish embassy made a triumphal entry into Rome. It included five allegoric carriages representing Portuguese control of the oceans and the crowning of Lisbon as well as ten other accompanying carriages. The elaborate display contributed to obtain from the pope the granting to the Portuguese crown of the monopoly of catholic presence in the Far East and higher ecclesiastic status for Lisbon. The courtier who headed and organized the parade was Rodrigo de Meneses, the Marquis of Fontes, who had served in the Portuguese army and had been trained as a military engineer in the Class of Fortification and Military Architecture. The presence of the Marquis in the papal city was not only destined to obtain religious privileges from the pope and transform Lisbon into new beacon of Christianity. Making use of his training as an engineer he organized a major operation also for transferring Rome's monumentality to Lisbon. Through his actions and with the crucial contribute of Italian architects and artists, the king D. João V collected at the royal palace in Lisbon an authentic museum of Rome's architecture, which included models, drawings, paintings, and sculptures.[39]

The building of a new royal palace complex in Lisbon able to rival the Roman example constituted one of the most emblematic projects, but it never materialized. Instead, for reasons not totally clear, a gigantic new convent and royal residence was built in Mafra, some 40 km to the north of Lisbon.[40] Its construction involved some 50,000 workers and constituted practical school for all the major construction endeavors that followed. According to historians of art and architecture, this Mafra experience fixed the image of the absolutist monarchy bringing together the lessons of the Italian baroque learned in Rome with the austere tradition of Portuguese engineers and their military construction techniques.[41] In 1718, the engineer Manuel da Maia presented the plans for a new road connecting Lisbon to Mafra so courtiers could travel between the two cities in their carriages. This baroque obsession with the mobility of noble carriages also justified the demolition of Lisbon's medieval gates as well as the opening of a new boulevard along the river bank. This new major Lisbon thoroughfare connected the royal palace in the center of the city to the new palaces to the west built by Portuguese nobles enriched by Brazilian commodities and who proudly traversed the new artery in their luxurious carriages.[42]

While the western parts of the city displayed the new baroque monumentality of the capital of the empire of D. João V built on the back of African slaves in Brazilian mines, the popular areas to the east were left out of the field of action of the king's engineers. The most striking example of this neglect of Lisbon eastern more plebeian neighborhoods is probably

Rome:' Dom João V of Portugal and relations between Rome and Lisbon" in *The Age of the Baroque in Portugal*, Ed., Jay Levenson (Washington, 1993), pp. 49–80.

[39] W. Rossa, "A imagem ribeirinha de Lisboa: Alegoria de uma estética urbana barroca e instrumento de propaganda para o império," *Proc. of the Meeting Barroco Iberoamericano: Territorio, Arte, Espaço y Sociedad*, (Seville: Ediciones Giralda), pp. 1314–1343.

[40] António Filipe Pimentel, *Arquitectura e Poder—o Real Edifício de Mafra*, (Coimbra, 1992).

[41] Paulo Varela Gomes, *A Cultura Arquitectónica e Artística em Portugal no Séc. XVIII*, (Lisboa, 1988).

[42] W. Rossa, "A imagem ribeirinha."

CONSTRUÇÃO DO CONVENTO DE MAFRA

Figure 2.9: **Building of the Convent of Mafra**.
Watercolor by the Portuguese artist Roque Gameiro for the 1917 album *Quadros da História de Portugal* (Pictures of the History of Portugal).
The building of Convent of Mafra started in 1717 and constituted a practical school for all the major projects that followed under the absolutist rule of D. João V. This watercolor is part of an album produced in the early twentieth century that offered a nationalistic history of the country in eight historical cycles from the origins of Portugal to the Republican period. The building of the convent was part of the fifth cycle, the "period of magnificence." *Baltasar and Blimunda* (Portuguese: *Memorial do Convento*, 1982), the historical novel by the Nobel Prize-winning Portuguese author José Saramago, describes in detail the challenges involved in the building of the convent, masterfully intertwining its material and spiritual dimensions.
© Public domain. Pictures available at www.roquegameiro.org - A Tribo dos pincéis.

the building of a monumental new aqueduct started in 1731 for which the engineers Manuel Azevedo Fortes and Manuel da Maia made the topographical work.[43] Maia would later assume the direction of the works of an infrastructure that ignored the water supply needs of the old city and focused instead on the fancier neighborhoods of the west and the needs of the new noble and royal palaces in the area. And again, engineers combined utility and aesthetics: Following the Roman example, monumental fountains were placed at the center of new squares offered to the city dwellers by their absolute monarch. These were designed as baroque scenarios, erected

---

[43] Irisalva Moita, Ed., *D. João V e o abastecimento de Água a Lisboa*, (Lisboa: Câmara Municipal, 1990).

Figure 2.10: **View of the grand Aqueduct over the Valley of Alcântara near Lisbon by Alexandre-Jean Noël (engraving by J. Wells), c.1790**.
The expansion of the state infrastructural power in the 18th century materialized in the territory through buildings that came out first and foremost from the drawing tables of military engineers. The Águas Livres Aqueduct is one of the finest examples. Its main course covers 18 km, but the whole network of canals extends through nearly 58 km. Construction started in 1731 under the direction of Portuguese engineers, (including Manuel da Maia, Manuel de Azevedo Fortes, Eugénio dos Santos and Carlos Mardel). It survived the 1755 earthquake.
© Public domain. Collection of the Museu da Cidade de Lisboa.

to stage the magnificence of D. João V. Significantly, one of the arches of the aqueduct lost the purity of lines of the engineers' drawing and was refurbished as monumental city gate, a triumphal arch celebrating the Magnanimous king.

It was thus only appropriate that Manuel de Azevedo Fortes, as chief engineer of the kingdom, had his portrait painted by Pierre-Antoine Quillard, the French artist hired by D. João V. While celebrated for the extravagant parties he produced for the amusement of the court, Quillard's fame as an artist was also due to his several portraits of the royal circle.[44] The most distinguished of Portuguese engineers was not only an accomplished scholar promoter of Cartesian ideas and distinguished mathematician; a translator of the most up-to-date European methods for building a fortification; a member of the military able to take decisions about the

---

[44] António Filipe Pimentel, "Os pintores de D. João V e a invenção do retrato de corte," *Revista de História de Arte*, 5 (2008), pp. 133–151.

best location for a defense structure guaranteeing the stability of the Empire; a producer of maps drawn according to standardized methods and that attested the historical legitimacy of the kingdom's posessions; a surveyor responsible for the topographical work demanded by new major infrastructures. He was also a courtier, who contributed through his work to the radiance of the monarch and thus deserved to have his portrait painted by one of the court's most cherished artists.

## 2.4    ENGINEERS AND THE LISBON EARTHQUAKE (1755)

The skills of the engineer embodied by Azevedo Fortes—mathematical training, map drawing, infrastructure building (military and civil), courtier talent—were all put at work after the catastrophic earthquake and tsunami that destroyed Lisbon in 1755. Manuel da Maia would be Fortes' successor as chief engineer of the kingdom. He was certainly no less erudite than Fortes, being responsible for no less than the state archives, saving millions of documents from the cataclysm and enabling the continuity of the bureaucratic memory of the Portuguese state. Again, these entanglements between engineering and history would deserve fuller exploration. But Maia is better known for working closely with the powerful Marquis of Pombal (Sebastião José de Carvalho e Melo, Count of Oeiras) to rebuild the capital of the empire in the aftermath of a disaster that killed some 30,000 people and caught the attention of every intellectual in Europe, including Voltaire, Rousseau, and Kant. Maia needed no more than 33 days to present to the Marquis, then secretary of state of the interior who controlled the state apparatus during the reign of D. José I (1750–1777), "the guidelines of what would become the plan for the renovation of the central area of the city."[45] He would then nominate Eugénio dos Santos and Carlos Mardel for the undertaking of the plan, both military engineers with vast previous experience in building fortifications, designing palaces, and managing the works of the monumental Lisbon aqueduct.[46]

In the newly established Casa do Risco (Drawing House), an adhoc institutional setting common in previous major building endeavors, engineers produced a new orthogonal urban layout for the capital city with a clear hierarchy of streets that guaranteed the unimpeded flow of people, carriages, air, light and waste water. Also from the Drawing House came detailed architectural plans for individual buildings imposing a modular typology, standardized building components, and new construction methods resistant to seismic events and fire. Twenty-five years after the earthquake, the number of dwellers in the affected area was already bigger than the original one in 1755. The characteristic new austere geometrical image of the rebuilt capital was inseparable from the engineers' knowhow previously tested in those frontier military outposts across the empire, namely in the Amazon.

---

[45] W. Rossa, "Dissertação sobre reforma e renovação na cultura do Pombalismo," in *O Terramoto de 1755: Impactos Históricos*, Ed., A. C. Araujo, J. L. Cardoso, N. G. Monteiro, J. V. Serrao, and W. Rossa (Lisboa: Livros Horizontes, 2007), p. 383.
[46] Christovam Ayres, *Manuel da Maya e os Engenheiros Militares Portugueses no Terremoto de 1755*, (Lisboa, 1910).

Figure 2.11: **Portrait of Carlos Mardel by unidentified painter**.
Carlos Mardel was a Hungarian-Portuguese military officer, engineer, and architect. Mardel was one of the engineers involved in the building of the Águas Livres aqueduct, also responsible for the design of the new urban layout of Lisbon after the catastrophic earthquake of 1755.
© Public domain.

Figure 2.12: **Portrait of Eugénio dos Santos by unidentified painter**. Eugénio dos Santos was a Portuguese architect and military engineer, who together with Carlos Mardel undertook the plan for the reconstruction of Lisbon after the 1755 earthquake. These engineers produced a new orthogonal urban layout with a clear hierarchy of streets. Engineers also detailed plans for individual buildings imposing a modular typology and standardized building components that contributed to the new austere geometrical image of the rebuilt capital.
© Wikimedia Commons.

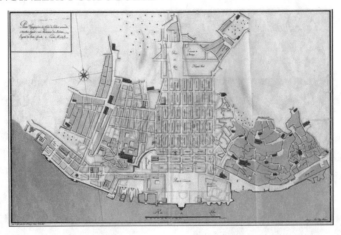

Figure 2.13: **Plan of the reconstruction of Lisbon after the 1755 earthquake**.
The entire "Baixa" quarter (downtown Lisbon, near the river Tagus, the most affected area)
was razed so that it would be possible to "lay out new streets without restraint." The engineers
Carlos Mardel and Eugénio dos Santos, whose signatures are in this plan, took the responsibility
of designing a city based on Enlightenment principles materialized in a geometrical grid of
parallel and perpendicular roads and streets. Their previous experience in building fortifications,
organizing sieges, designing palaces and managing the works of the aqueduct now served the
(re)invention of Lisbon.
© Public domain. Collection of the Museu da Cidade de Lisboa.

Architects, trained in the tradition of the beaux arts, would later criticize the dullness of
an aesthetics they judged more appropriated to military barracks than for emblematic buildings
of the kingdom's capital, but engineers would prove nevertheless once again their attention and
responsiveness to the demands of the court. The newly rebuilt central square of Lisbon, the
Commerce Square—a vast rectangle opened to the majestic views of the river Tagus—combined
geometrical perspectives and repetition of standard elements in windows and arches with a new
major bronze statue of the monarch first designed by Eugénio dos Santos, the engineer in charge
of much of the reconstruction effort. The new Lisbon, while (re)producing the impression of
functionalism and improvization of the military frontier cities of northern Brazil, offered as
well the appropriate monumentality for the deployment of a court society.[47]

The experience in the reconstruction of Lisbon was soon expanded into the rest of the
country. The expansion of the state infrastructural power and its growing intervention in the
country's economy in the second half of the 18th century materialized in the territory through

[47] Walter Rossa, "Do plano de 1755–1758 para a baixa-chiado," *Monumentos*, 21 (2004), pp. 22–43.

buildings that came out first and foremost from the drawing tables of military engineers:[48] The urban renewal of Porto and Coimbra, the second and third cities of the country; the planning of new cities from scratch (Vila Real de Santo António); new royal factories and industrial neighborhoods (Amoreiras); hospitals (Caldas); army barracks; warehouses...[49] These were the exemplary constructions mobilized at the end of the century by the military engineer Carvalho Negreiros, son and grandson of military engineers to defend his corporation from the attacks of architects trained in the neoclassical tradition in France. For Negreiros there was no difference between architecture and engineering, the first constituting the last stage of an engineer's educational journey. This started with trigonometry, followed by mechanics, hydraulics, and hydrostatics, continued by building, and culminating in urban planning and architecture. This clear educational structure constituted the basis of the different institutional forms of engineering education of the period.

The reign of D. José I (1750–1777) was particularly rich concerning educational reforms, namely in making the University of Coimbra, the major higher education institution of the country, more responsive to the Marquis of Pombal economic policies marked by state centralization and industrial promotion.[50] Nevertheless, the training of Portuguese engineers remained almost undisturbed. In fact, only minor changes were made: a course on drawing and architecture was introduced as part of the degree in Mathematics at the University of Coimbra, which became equivalent to a degree in engineering; a course in military architecture was also taught at the new *Real Colégio dos Nobres* (Royal College of Nobles, 1761–1837) but with very little success. However, reforms of the Portuguese army directed by the German officer Count Lippe (Friedrich Wilhelm Ernst zu Schaumburg-Lippe) proved very influential concerning the professional prestige and *status* of military engineers. In the *Military Observations* the count sent to Marquis of Pombal in 1764 discussing the reform of the Portuguese army at the image of the Prussian example, he called attention to the role of engineers in modern warfare and included the proposal for an independent corps of engineers detached from infantry. Engineers

---

[48] Jorge M. Pedreira, *Estrutura Industrial e Mercado Colonial: Portugal e Brasil, 1780–1830*, (Lisbon: DIFEL, 1994); Nuno Luis Madureira, *Mercados e Privilégios: A Indústria Portuguesa Entre 1750 e 1834*, (Lisbon: Estampa, 1997); N. M. S. Salgueiro, Fabricar o território: Os equipamentos do sistema produtivo Português, (1670–1807), MA dissertation, Universidade de Coimbra, 2006.

[49] Paulo Varela Gomes, "Jornada pelo Tejo: Costa e Silva, Carvalho Negreiros e a cidade pós-pombalina," *Monumentos*, 21 (2004), pp. 132–137.

[50] Although the first evidence of a growing interest for a new cultural attitude vis-à-vis science and technology can be found during the reign of King João V, it was during the reign of his son, King José I, that this new mentality became a government affair. The Marquis of Pombal, the powerful prime minister of José I and a man of the enlightenment, implemented an educational reform, deeply shaped by utilitarianism and rationalism, which aimed at changing the economic and social situation of the country. The Marquis' prime objective was to develop a national scientific and technical elite, supported by a new educational system. The main aspects of his educational reform were the creation of the Royal College of Nobles of Lisbon (1761–1837) and the reform of the University of Coimbra (1772). In the Royal College of Nobles, Pombal introduced for the first time in Portugal a school curriculum including courses in science, mathematics and experimental physical sciences. Ten years after its creation, scientific training at the Royal College came to an end. The Marquis of Pombal' efforts toward the implementation of a new kind of education turned then to the University of Coimbra. He split the University into six Faculties (Theology, Cannon Law, Civil Law, Philosophy, Medicine and Mathematics, the last two newly created) and established a Chemical Laboratory, a Botanical Garden, a Cabinet of Physics, and an Observatory.

would nevertheless have to wait thirty-six years more to have their own corps, its formation being dependent on reforming their training.

In 1779, the *Real Academia da Marinha* (Royal Academy of the Navy) formally replaced the Class of Fortification and Military Architecture, which, as we saw, had trained Portuguese engineers since 1647. The Royal Academy of the Navy aimed at changing the alleged still predominantly empirical characteristics of the education of engineers, by emphasizing its scientific dimension: students had to attend compulsorily classes of Mathematics and Physics (static, dynamics, hydrostatics, hydraulic, and optics). Only after having passed successfully their exams in these courses could students attend courses on fortification and military and civil architecture.

Eleven years after (1790), the Royal Academy of the Navy was complemented by the *Real Academia de Fortificação, Artilharia e Desenho* (Royal Academy of Fortification Artillery and Drawing). Only after completing the Mathematics and Physics courses at the Royal Academy of the Navy could students enter this so-called "first true engineering course" opened on January 20, 1790; this new four years course included both a strong theoretical *curriculum*, still largely based on Fortes' 60-year old textbook, and practical classes in the field. The continued use of Fortes' *The Portuguese Engineer* suggests that changes in institutional names didn't translate into radical different practices in engineering education. Nevertheless, the centralization effort was obvious: all those previous provincial military academies in border towns were shut down and the Royal Academy of Fortification centralized the teaching of engineering. Concerning the army career, officers that concluded the course were chosen in preference to those who had no formal education.

The *curriculum* of this course is quite revelatory of the dual profile that constituted the mark of engineering training in the country: the first three years were clearly military driven, covering topics such as the general rules of fortification, the general theory of attack and defense, artillery and mines; the last year was dedicated to civil engineering, focusing on "civil architecture, stone and wood cutting, budgeting, and everything related to materials as well as the explanation of the best methods to build roads(…) they had to learn hydraulics and the building of bridges, channels, ports and dams."[51] Many new influential books that arrived at the Academy's library from Europe, such as Jean Rodolphe Perronet's treaty on bridge building were indeed related to civil engineering.[52] The students coming out of the Academy of Fortification would join military ranks, but they were also increasingly called to enroll in the building of public works, namely port facilities or new major roads such as the one connecting Lisbon and Porto, the two main Portuguese cities. In fact, the promoters of the Academy of Fortification were explicit about the importance of bringing to the country the latest European developments in both military and civil engineering, more specifically those identified with the French *École du Corps du Génie* (Military School) and the *École des Ponts et Chaussées* (Civil Engineering School). The Portuguese ambassador in Paris, after hiring in June 1789 the young French engineer Joseph Aufddiener

[51] Decree, January 2, 1790.

[52] Carlos H. M. R. Martins, O programa de obras públicas para o território de Portugal continental, 1789–1809, Ph.D. dissertation, University of Coimbra, 2014, p. 635.

from the Parisian school of civil engineering to direct the new major public works planned for Portugal, mentions in his correspondence with the minister of the kingdom that the Frenchman would be carrying with him information concerning the organization of engineering schools in his country. He candidly acknowledges that this would contribute "to found there [in Portugal] one [school] imitating the French one [the school of Civil Engineering], which is the best in whole Europe."[53]

The structure of the four-year course at the Academy of Fortification preceded by preparatory classes at the Academy of the Navy certainly owed much to the French model, the most impressive engineering training system in the continent at the end of the 18th century. But after the preceding pages, there is no doubt that the new Academy of Fortification was mostly built on the previous Portuguese experience of the Class of Fortification. And it is important to retain that in contrast to the French example, in which civil and military schools were separated, in Portugal they would be kept together. The same year the Academy was formed (1790), and as part of the same initiative, the Royal Corps of Engineers was finally institutionalized as an independent branch of the army. It was formed by 88 officers: 3 generals, 10 coronels, 13 lieutenant coronels, 15 majors, 30 three captains, 14 auxiliaries. While some of the members of the Corps had been trained in the Academy of the Navy or in the old Class of Fortification, many had had only a more rudimentary training in the regional academies or had been self-taught. The new Academy of Fortification should now be the first supplier of officers to the Corps, the new rank of second lieutenant being given to those that completed its four-year course. Scientific training increasingly became the watermark for those willing to join the Corps. And while this was a military structure it also integrated those engineers exclusively employed on the building of public works and who had no military experience.

The dual profile—civil and military remained a characteristic of engineering training until very late in Portugal. Engineers were envisaged has having a double identity, both military and civil. In fact, discussions on the nature and aim of the Portuguese system of engineering training would continue all along the 19th century, with heated debates concerning its methodological and epistemological premises, as well as its role in serving the nation.

---

[53] Letter from Vicente de Sousa Coutinho to Luis Pinto de Sousa Coutinho (June 11, 1789), quoted by Carlos H. M. R. Martins, *O Programa de Obras Públicas*, p. 635.

CHAPTER 3

# Engineering the Liberal State

The official imprint of the role of engineers as constitutive elements of the Portuguese state arrived in 1790 with the founding of the Royal Corps of Engineers. Engineers were finally recognized as an autonomous group of experts inside the Portuguese military with a specific training based on a unique mix of theoretical and practical knowledge. Nevertheless, the royal decree defining the statutes of the corps was only published in 1812 when the Portuguese court was already exiled in Brazil after fleeing the Napoleonic invasions (1807, 1809, and 1810).[1]

The war with the French, the exile of the court in Rio de Janeiro (Brazil), the independence of Brazil (1822), and the violent civil war (1828–1834) that followed between absolutists (supporters of an absolutist monarchy) and liberals (supporters of a constitutional monarchy) meant great instability in the country and an abrupt discontinuity in state structures.[2] Engineers demonstrated their importance during the civil war in organizing sieges or building up defense lines, but they did it on both sides. When absolutists took control of the country, many liberal engineers went into exile in France and England, only to return in 1832 as part of the liberal military offensive. After liberal victory in 1834, engineers supporting the cause of the absolutist king were reincorporated in the new army but at lower rank.[3] The comradeship forged during war time among engineers of the liberal army would prove essential in the upcoming years in which they would take political control of the country. And if in the first decades of the century engineers split according to political ideological lines, from the midcentury onwards they themselves developed a new political ideology which would dominate Portugal until the end of the 19th century.

---

[1] Decree, January 12, 1812.

[2] Valentim Alexandre, *Os Sentidos do Império: Questão Nacional e Questão Colonial na Crise do Antigo Regime Português*, (Porto: Ed. Afrontamento, 1993); Jorge Pedreira, "From growth to collapse: Portugal, Brazil, and the breakdown of the old colonial system," *Hispanic American Historical Review*, 80, no. 4 (2000), pp. 839–864; Nuno Monteiro, *Elites e Poder entre o Antigo Regime e o Liberalismo*, (Lisbon: ICS, 2003); Vasco Pulido Valente, *Os Devoristas: A Revolução Liberal, 1834–1836*, (Lisbon: Quetzal, 1993).

[3] See the biographies of engineers provided in Cristóvão Ayres de Magalhães Sepúlveda, *História Orgânica e Política do Exército Português*, vol. VII and vol. VIII, (Coimbra: Imprensa da Universidade, 1919).

Figure 3.1: **The Jesuit novitiate of Cotovia (1603-1759), later Royal College of Nobles (1761-1837), Royal Navy Academy (1779-1837), Royal Academy of Fortification (1790-1837), Army School (1837-1843) and Polytechnic School (1837-1911).** This iconic building housed initially a Jesuit novitiate, becoming the main center for the training of Portuguese engineers from the late eighteenth century onwards until it was destroyed by a fire in 1843.
© Public domain.

## 3.1   THE MAKING OF THE LIBERAL ENGINEER: THE ARMY SCHOOL AND THE LISBON POLYTECHNIC SCHOOL

Only in 1834 were liberals able to consolidate their control of the country. Less than one year after, they formed a Council of Public Education aimed at reforming the entire educational structure of the country, demonstrating how essential the issue was for politicians invested in the formation of a learned public sphere able to contribute to collective decisions and thus consolidate the new liberal regime. In the first decades of the 19th century the Royal Academy of Fortification Artillery and Drawing remained the only school offering formal training in engineering. From 1790–1836, no less than 2736 students registered, but only 181 got their degree, i.e., a ratio of less than 4 engineers per year.[4] Such numbers are an eloquent proof of the country's turmoil and how it profoundly affected the engineering profession. The liberal regime, as we shall see, would indeed produce a new kind of engineer that although based on the previous experience of military engineering of the 17th and 18th centuries would dramatically redefine his positioning in relation to the state.

---

[4] From the 2,736 students enrolled only 462 succeeded in their final exams, i.e., around 17%, Sepulveda, *Historia Organica*, vol. VI-Provas, 152.

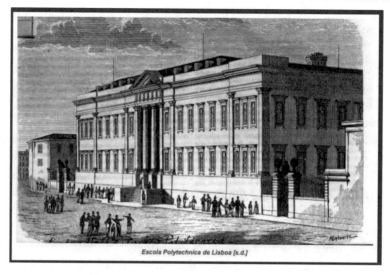

*Escola Polytechnica de Lisboa [s.d.]*

Figure 3.2: **Lisbon Polytechnic School**. The Polytechnic School building inaugurated in 1863 replaced the previous facilities destroyed by fire in 1843. The Lisbon Polytechnic school became the main scientific institution in Lisbon, training not only engineers but also future pharmacists or surgeons. Engineering education was nevertheless its core business. The Polytechnic School was under the Ministry of War control until 1859 and the large majority of its students became members of the military, completing their education at the Army School.
© Public domain.

Anticipating an extensive set of reforms, the liberal government through the Council of Public Education had tried in 1835 to end the exclusive control in scientific higher education of the oldest Portuguese university—the University of Coimbra[5]—commonly identified as an institution serving the old regime. The new liberal elites, based on foreign experiences learned while in exile in France and England during the absolutist rule, put forward an Instituto de Ciências Físicas e Matemáticas (Institute of Physics and Mathematics), which included a School of Civil Engineering. However, the Institute never opened, mainly because of the political turmoil between different liberal factions that continued to mark this period and the effective opposition from the faculty of the University of Coimbra.[6] But in 1836, such resistance was finally overcame by creating new scientific and technical institutions directly dependent of the Min-

[5] In 1290 the Portuguese King D. Dinis published the royal charter *Scientiae Thesaurus Mirabilis* announcing the establishment in Lisbon of the *Studium Generale* (Estudo Geral). This first university went through a number of relocations until it was moved permanently to Coimbra in 1537, being the oldest university of Portugal and one of the oldest universities in continuous operation in the world. Coimbra is located in the central region of Portugal, being the fourth-largest urban center in the country.

[6] This institute would encompass the following special schools: School of Civil Engineering, School of Military Engineering, Navy School, School of Pilots, School of Trade.

istry of War. The same military who had guaranteed the victory of liberalism over absolutism understood scientific and technical training as the key to perpetuate their prominent role in the new regime.

The Escola do Exército (Army School)[7] replaced the Royal Academy of Fortification Artillery and Drawing and two new polytechnic schools were created: the Escola Politécnica (Polytechnic School) in Lisbon and the Academia Politécnica (Polytechnic Academy) in Porto (this one under the control of the Ministry of Interior—Ministério do Reino).[8] This set of institutions would dominate Portuguese engineering training until the beginning of the 20th century. The continuities with the previous institutional arrangement were obvious: The Army School and the Lisbon Polytechnic School had taken over the premises of the Royal Academy of Fortification Artillery and Drawing and of the Royal Academy of the Navy. While before, students had to attend the Academy of the Navy to be admitted at the Academy of Fortification, they now had to complete four years at the Polytechnic School to be considered for admission at the Army School where they could aspire to acquire the tittle of engineer.

Two engineering courses were taught at the Army School, one on military engineering and a second one on civil engineering.[9] The fact that a civil engineering course was taught at the Army School may seem, at first glance, awkward, but, again, it just extends the 18th-century trend of training engineers for, as stated in the official bulletin of the government, "two different kind of professional tasks: civil and military. We will have still to wait for a long time until a new class of civil servants, i.e., Civil Engineers, may relieve them from the first kind of tasks."[10] Once students were admitted at the Army School, and independently of having opted for civil or military engineering degree, they automatically became part of the Portuguese military, ranked as student-second lieutenant (*alferes-aluno*), and were given a uniform. From 1837–1890, 305 students completed the military engineering course and 114 the civil engineering one.

The military engineering degree required three years of study and seven courses divided into sets of lectures; the degree in civil engineering at the same Army School was less demanding and took only two years to complete and was built on the military course, using only part of its curriculum, namely: the 1st lecture of the 6th course—General rules on the building of military bridges; the 5th lecture of the 2nd course—Building materials; 4th course—Stability and mechanics applied to machines and hydraulics (which included an appendix on the building of suspension bridges); 5th course—Civil architecture (which included railways and roads); 6th course—Topography and drawing; 7th course—English.

---

[7] Decree, January 12, 1837.

[8] Decrees, January 11 and 13, 1837.

[9] Besides these courses three other courses devoted to military training were taught.

[10] *Diário do Governo*, no. 239 (1839), p. 1476.

The textbooks used at the Army School came mostly from the French *École des Ponts et Chaussées*[11] and *the École Centrale*,[12] although with some adaptations to Portuguese specificities. Portuguese engineers should be proficient both in French and in English to be able to follow the two main international references of their area of expertise. While in the previous Academy of Fortification students still had Azevedo Fortes' treaty as major source, at the Army school they relied on the recent knowledge produced in the most renowned Engineering schools in Europe. Another major difference with the previous training at the Academy was the increased amount of time dedicated to civil engineering subjects (like railways, roads, or harbors) at the expense of military ones. This growing emphasis on civil engineering would constitute a defining element of Portuguese engineers' identity in the 19th century. This said, civil engineering, continued to be considered part of the military training, having no autonomous status. Public works kept being carried out by military engineers, who were part of the all-embracing "technical services," which were in charge of "the defense of the country, civil works, roads, geological and other surveys, draining, improvement of ports and the supervision and management of arsenals."[13]

In spite of the many continuities with the previous system of engineering training, novelties are more salient when looking at the Lisbon Polytechnic School, the institution designed to prepare students to enter the Army School.[14] At the Polytechnic, students were educated during four years in the different sciences which were understood as the basis for the applied knowledge taught at Army School. Such division between fundamental and applied knowledge, translated institutionally into a preparatory school and an application one, had been already the basis for the previous distinction between the Academy of the Navy in which students learned mathematics and physics, and the Academy of Fortification, dedicated to the teaching of the specific skills of the engineering profession. And in fact, the three professors who lectured at the Academy of the Navy transited to the new Polytechnic School. But in addition to these three, ten other professors had to be hired to the new institution. This was justified because the body of scientific knowledge was much more comprehensive at the Polytechnic than at the Academy of the Navy. In addition to mathematics,[15] drawing, astronomy, geodesy, physics, and mechanics, students at the Polytechnic also had to attend courses on botany, geology, chemistry, mineral-

[11] The École des Ponts et Chaussées is the main European reference as far as engineering training in civil engineering is concerned. It was founded in 1747 by Daniel-Charles Trudaine under the name of *École Royale des Ponts et Chaussées*, (Royal School of Roads and Bridges) and it is the oldest school of civil engineering in the world and one of the most prestigious French engineering schools.

[12] The École Centrale Paris was founded in 1829 by Alphonse Lavallée, who became its first president, and three scientists who became founding associates. Its aim was to educate multidisciplinary engineers for the emerging industrial sector in France. At first a private initiative, the institution was offered to the French state in 1857 by Lavallée.

[13] *Diário do Governo*, no. 251 (1859), p. 1361.

[14] The Polytechnic Academy (Oporto) had a civil engineering degree, but its curriculum was much closer to a general scientific course then to an engineering one. In spite of aspiring to be considered as the Portuguese version of the French École Centrale, it never reached the same prestige of the Army School and of the Lisbon Polytechnic.

[15] The central role of mathematics in the training of modern engineers has been largely discussed by historians of science and technology. See Ken Alder, *Engineering the Revolution: Arms and Enlightenment in France, 1763–1815*, (Princeton, NJ: Princeton University Press, 1997); Antoine Picon, *L'invention de L'ingénieur Moderne: L'École des Ponts et Chaussées, 1747–1851*, (Paris: Presses de l'École des Ponts et Chaussées, 1992).

ogy, metallurgy, political economy, and law, the last two being critical to students who would serve as civil servants.

The explicit model for the Lisbon Polytechnic School preparing students through a wide-ranging scientific education for engineering application schools came again from France. The Parisian *École Polytechnique* founded in 1794 counted with the most distinguished French scientists (Laplace, Fourier, d'Alembert, Monge, Arago, etc.) to educate the students who would later enter the prestigious Artillery School, Civil Engineering School, and the School of Mines. In Portugal, the lack of institutional differentiation meant that the Army School alone assumed the role of culprit of technical education, but the relation between Polytechnic training and application school is the same as in the French case. It is of course no coincidence that the first director of the Lisbon Polytechnic School, José Feliciano da Silva Costa, had studied at the Parisian *École des Ponts et Chaussées* while exiled in Paris from 1824–1832 and that he made himself known among the new Portuguese liberal leadership for its role in the civil war.[16] And if 1837, the year of the creation of the Lisbon Polytechnic and the Army School, might seem as a big temporal gap in relation to the French 1794 inauguration of the *Polytechnique*, the timing was nevertheless well in tune with what was going on in engineering education across the rest of Europe. In Greece, Belgium, and Spain, the establishment of liberal monarchies in 1830s also meant the founding or reform of engineering schools following closely the Parisian model of producing engineers that had come out of the French Revolution. In all these cases, as in Portugal, it seemed impossible to imagine a new liberal society without rethinking what it meant to be an engineer.

The fact that the Lisbon Polytechnic school became the main scientific institution in Lisbon, one in which students training to become pharmacists or surgeons attended its chemistry or botany classes, should not distract us from the fact that engineering education was always its core business.[17] The school offered five courses, four of them preparing students for the different branches of the military (military officers, military and civil engineers, artillery officers, navy officers and engineers) and one general course.[18] Also, the Polytechnic shared with the Army School its administrative structure, procedures for hiring professors, and evaluation methods.[19] While in 1859 the school earned its civil status being controlled afterward by the Ministry of Interior instead of the Ministry of War, in the early 1880s practically all the students that completed its courses became members of the military.[20]

---

[16] Cristóvão Ayres de Magalhães Sepúlveda, *História Orgânica e Política do Exército Português*, vol. VIII (Coimbra: Imprensa da Universidade, 1919), p. 572.

[17] Revisionist historiography on the Polytechnic School has tended to downsize the overwhelming military dimension of the Polytechnic. See namely, José Luís Carolino and Teresa Salomé Mota, "Introduction," *HoST* 7, (2013), pp. 9–12; Luís Miguel Carolino, Teresa Salomé Mota, and Dulce Figueiredo, "The Portuguese polytechnicians of the "long 19th century:" Technical expertise, military aspirations, and political disenchantment. A preliminary study," *HoST* 7, (Spring 2013)/Special Issue The Polytechnic Experience in the 19th Century Iberian Peninsula: pp. 52–66.

[18] Decree, Janeiro 11, 1837.

[19] Marta Macedo, *Projectar e Construir a Nação*, (Lisboa:ICS, 2012), p. 55.

[20] José Silvestre Ribeiro, *Historia dos Estabelecimentos Scientificos Litterarios e Artisticos de Portugal nos Successsivos Reinados da Monarchia*, (Lisboa: Academia Real das Sciências, 1871–1914), vol. 12, p. 290.

Figure 3.3: **Lecture Theater of the Laboratorio Chimico, Lisbon Polytechnic School, 1891.** The explicit model of the Lisbon Polytechnic School was the Parisian École Polytechnique training engineering students through a wide-ranging scientific curriculum who would later complete their education in application schools. The chemistry laboratory of the Lisbon institution was an eloquent example of the basic training necessary to enter the Army School. © Arquivo Histórico do MCUL.

A catastrophic fire in spring of 1843 destroyed the building in Lisbon that housed the Polytechnic and the Army Schools, and which had also housed the Fortification and Navy academies in the previous decades. While the Army School would be given already in 1850 new premises in a former Royal Palace (Bemposta), the director of the Polytechnic, José Feliciano da Costa Silva, would design an ambitious new building for his school. The Polytechnic new monumental building was inaugurated in 1863 and constituted one of the urban landmarks of 19th-century Lisbon.[21] The students who first crossed its austere neoclassical façade with tall columns and triangular pediment found traditional classrooms where they would learn the differential calculus developed by French *Polytechniciens* (Laplace, Fourier, etc.) and drawing rooms for the study of Descriptive Geometry as elaborated by Gaspard Monge also at the Parisian *Polytechnique*. But later they also had at their disposal more impressive spaces such as a museum of Natural History and a Chemistry laboratory. If students left the building through its backdoor they could wander along the paths of the botanical garden (opened in 1873) as well as wander the skies by entering the astronomic observatory (opened in 1875). The building was no less than

[21] Tiago Saraiva, *Ciencia y Ciudad*.

the materialization of the positivist order of knowledge developed by August Comte (himself a *Polytechnicien*) in his *Course of Positivist Philosophy* published between 1830 and 1842.[22]

As mentioned, law and political economy were also part of the curriculum at the Polytechnic school, which might sound unexpected for contemporary engineers. This deserves explanation. Comte's positivist order started with mathematics at its basis, unfolded into the natural sciences, and culminated in the social sciences. Knowledge was thus placed in an encyclopedic scale moving from the general to the particular, from the simple to the complex. Engineers learned physics, chemistry, and differential calculus to know how to intervene scientifically in the world. But this world was not just the natural world. Engineering education should also supply students with the skills to apply science in the organization of the whole society. While the study of administrative law taught them how to reorganize the state according to rational principles, political economy, as developed also by French authors (Jean Baptiste Say and Michel Chevalier), offered them the tools to manage the national economy.[23]

Political economy as taught at the Polytechnic made engineers indispensable elements of the new liberal order. They were the ones who planned the material infrastructure that enabled the circulation of material goods, the basis of social welfare and ultimately of individual freedom, according to the scientific laws of political economy. Or as Michel Chevalier, quoted by the Lisbon Polytechnic professor of political economy António de Oliveira Marreca, reminded, "material interests are the indispensable support and essential condition of freedom."[24] Without material improvements, without the communication networks designed by engineers the freedom values evoked by liberal thinkers and lawmakers were meaningless. Students learned at the Lisbon Polytechnic School from Chevalier's political economy textbook that transportation, credit, and technical education, or in an alternative formulation, circulation of goods, money and knowledge, constituted the three building blocks of the new industrial society. Engineers had to be learned in all three to reclaim their leadership position in this new social order.

## 3.2    MAKING CIVIL ENGINEERS IN PORTUGAL AND ABROAD

In 1848, Albino de Figueiredo, who started his career as professor at the Navy academy and became later the professor of mechanics of the Polytechnic School and who also attended the *École des Ponts et Chaussées*, emphasized the importance of reinforcing the civil dimension of engineering courses. He insisted on the value of "consolidating a *corpus* of knowledge specific to civil engineers, a professional group without which any nation is deprived of its public works."[25] The question of how to train civil engineers and the balance between civil and military compo-

---

[22] Ana Luísa Janeira, *Sistemas Epistémicos e Ciências: Do Noviciado da Cotovia à Faculdade de Ciências da Universidade de Lisboa*, (Lisbon: INCM, 1987).

[23] Marta Macedo, *Projectar e Construir*, pp. 121–131.

[24] Quoted by Marta Macedo, "Projectar e construir," p. 129.

[25] *A Epoca*, Tomo I, (1848), p. 397.

nents would remain in the public sphere during the entire 19th century. In 1854, Júlio Máximo de Oliveira Pimentel, Chemistry professor at the Polytechnic School as well as a member of the Parliament, submitted a project that aimed at converting part of the military training institutions into Scientific and Technical Professional Schools: the Army School (for cavalry and artillery officers and military engineers) and the Navy School (for navy officers and shipbuilders) would be kept as part of the military training, but a new Public Works School (for civil engineers, builders, architects, geographical and hydraulics engineers, and mining engineers) and an Industrial School (for mechanical, chemical, and metallurgy engineers and foremen) were to be created. The project was not approved.[26]

The constant push for increasing the civil component of engineering training was deeply connected with the experience abroad of Portuguese engineering faculty from the Polytechnic School and the Army School in other European engineering schools of application, namely the French ones. From 1825–1851, 19 Portuguese engineers studied at the *École des Ponts et Chaussées*. It was through such travels that French textbooks ended in the libraries of Portuguese institutions and became the main reference texts for the training of Portuguese engineers.[27] In 1859, the Parliament resumed the debate on the training of Portuguese engineers, urging the Ministry of Public Works to send at least three students every year to study abroad. The Parisian *École des Ponts et Chaussées*, the *École des Mines* together with the civil engineering school in Gand (Belgium), the mining schools in Freiberg (Germany), and Liège (Belgium) were considered the ideal scientific establishments to complete Portuguese engineering training. After a period abroad, students were expected to return "with the training required to fulfil the noble functions of an engineer and through useful work payback Portugal what the country had invested in them."[28]

The idea of sending students to study abroad was part of a long tradition. As mentioned before, Portugal was used to "buy" knowledge, science and technology in the European marketplace as an alternative to create the conditions for developing local expertise and skills. The notion of "payback" emphasizes, on the other hand, the obligation of using what was learned abroad at home. A considerable number of these "travels of learning" were made at the expenses of the Portuguese government and aimed at improving specific expertise concerning mostly public works. Between 1845 and 1897 some 20 of those travels took place to France, Belgium, the Netherlands, England, and, less frequently, to Spain. Both young and already reputed engineers

---

[26] The proposal of closing down the Polytechnic Academy of Oporto and other schools, and the minor role ascribed to the Lisbon Polytechnic generated strong opposition, in particular from the teaching staff of the Oporto Polytechnic, which addressed a petition to the Members of Parliament. The petition was later on published in the *Jornal da Associação Industrial Portuense*, (nos. 19 (pp. 296–304), 20 (pp. 312–320), and 21 (pp. 330–336), under the tittle, *Breve Memória sobre a Instrucção Publica Superior no Porto e nas Provincias do Norte, Offerecida aos Senhores Deputados da Nação Portugueza pelos Lentes da Academia Polytechnica.*

[27] Macedo, *Projectar*, p. 80.

[28] *Diário do Governo*, no. 251 (1859), p. 1361.

O grande engenheiro João Evangelista de Abreu e o seu elogio historico pelo sr. Luciano de Carvalho.

João Evangelista de Abreu

Figure 3.4: **João Evangelista de Abreu**. The engineer João Evangelista de Abreu is a fine example of the strong connections between Portuguese engineering education and French institutions. Being already responsible for courses on roads and railroads at the Portuguese Army School, he went to Paris in 1856 to complement his training at the *École de Ponts et Chaussées*. After Abreu returned to Portugal, he became part of the technical elite involved in the multiple infrastructure plans of the liberal state.

© Public domain. In *Revista Brasil-Portugal*, 135, 1 September 1904. Available online at http://hemerotecadigital.cm-lisboa.pt/OBRAS/BrasilPortugal/1904_1905/N135/N135_master/N135.pdf

committed to these technical pilgrimages that focused mainly on roads, canals—railways and ports—but which also included industrial settings.[29]

João Evangelista de Abreu is a fine example of this breed of engineers. He was selected to enroll in the *École de Ponts et Chaussées* in 1856.[30] He held a bachelor degree in Mathematics

[29] Although they were an exception, some engineers went abroad to study industrial topics. In 1873, the Companhia de Torres Vedras, a textile factory, asked the engineer Jaime Archer to study the most updated features of mechanical looms in Northern Europe in order to renew the factory's existent machinery; in 1894, the firm António Moreira Rato & Filhos sent the engineer Herculano Galhardo to several European factories in order to study the latest innovations concerning both knowledge and machinery for producing Portland cement, which were actually introduced in the factory. See Ana Cardoso Matos and Maria Paula Diogo, "Bringing it all back home: Portuguese engineers and their travels of learning (1850–1900)," *HoST 1*, (2007), pp. 155–181.

[30] Ana Cardoso de Matos and Maria Paula Diogo, "From the École des Ponts et Chaussées to Portuguese railways: the transfer of technological knowledge and practices," in *Railway Modernization: An Historical Perspective (19th–20th Centuries)*,

from the University of Coimbra. From 1852–1853 he taught Geometry and Mechanics applied to the Arts (i.e., industry) and Professions, at the Lisbon National High School, and from 1854–1856 he was responsible for the course on roads and railways at the Army School.[31] Going to the *École* offered him the opportunity to improve both his theoretical and practical training. Together with the more conventional classes, students at the French school were encouraged to have a close contact with ongoing engineering works, participating in field trips to important building sites to experience the "real" world of public works: in 1857, João Evangelista de Abreu went to Pau and Marseille. At the end of their field trip he wrote a report, in which he praised this kind of training; the following year he went on a new fieldtrip to Bordeaux to study the Gironde estuary, the new iron bridge over the Garonne that linked the railway lines of Orléans and the Midi, the procedures for fixing dunes and its economic results, and the network of water distribution.[32] Once João Evangelista de Abreu returned to Portugal he became part of the technical elite involved in the building of a national railway network, the most ambitious plan of the liberal state. He was appointed head of the 2nd inspection division for the railways; took on the task of directing the railway studies south of the River Tagus;[33] and directed the works both in the Eastern and Northern railway lines.[34]

## 3.3 PORTUGAL REGENERATED OR THE SAINT SIMONIAN ENGINEERS IN POWER

From Abreu's exemplary story it doesn't take much to make the case about the importance of circulation in general, and railways in particular, in defining Portuguese engineers' identity in the second half of the 19th century. In the newspaper O *Atheneu*, published in 1850 and 1851, engineers would explore such connection in detail. The social influence of engineers went beyond their educational role as professors at the Army School and at the Polytechnic School or as distinguished members of the Army. They were also main actors in the emerging Lisbon bourgeois public sphere, and could be found discussing politics in cafes or imparting scientific conferences in clubs. They were particularly active in the multiple newspapers that emerged in the Portuguese capital city, some of them, such as the *Atheneu*, the result of engineers' own initiative.[35] Albino de Figueiredo, the already mentioned professor of mechanics at the Polytechnic and another veteran of the civil war opposing absolutists and liberals, explained profusely in the pages of this newspaper how communication networks (roads, canals, railways, telegraph) constituted

Ed., Magda Pinheiro, (Lisboa: CEHCP-ISCTE-IUL, 2009, pp. 77–90; Maria Paula Diogo, "João Evangelista de Abreu—Biographie," *Annuaire de l'École des Ponts et Chaussées*, 2008.

[31] AHMOPCI, Personal file for João Evangelista de Abreu.

[32] "Relatório sobre o emprego do tempo de missão feita no fim do segundo ano do curso de Pontes e Calçadas pelo aluno João Evangelista de Abreu, dirigido ao excelentíssimo Senhor Visconde de Paiva, Ministro de sua Magestade Fidelíssima em Paris" in *Boletim do MOPCI*, no. 8, August, 1860, p. 112.

[33] *Boletim do MOPCI*, no. 11, November 1866, p. 330.

[34] Frederico Abragão, *Caminhos de Ferro Portugueses. Esboço da sua História*, (Lisboa, Oficinas Gráficas da Gazeta dos Caminhos de Ferro, 1956), pp. 244–247.

[35] Saraiva, *Ciencia y Ciudad:* Macedo, *Projectar e Construir*, 159-164.

the binding elements of the nation, how they guaranteed its wealth and independence through the circulation of industrial goods. The efficiency of the network of communications allegedly enhanced the circulation of goods, thus supporting national industry. According to Figueiredo's proposals, the best engineers in the country would form a Council of Public Works and should be trusted with the design of a general plan of communications.[36] The construction work would be overseen by an extended body of well-trained engineers, and it would be conveniently funded according to judiciously prepared budgets. Albino de Figueiredo's French peers would certainly be proud of the Portuguese engineer combination of transportation infrastructure, technical education, and financial considerations. These three elements summarized Michel Chevalier's political economy, the trinity of the new engineering religion of Saint Simonianism.

Michel de Chevalier had in fact been invested as cardinal of this new religion, and would become editor in 1830 of its mouthpiece *Le Globe*, which publicized the technological utopia preached by Henri de Saint Simon. Saint-Simon's "New Christianism" announced a future era of peace and prosperity in which machines, communications, and credit would abolish social antagonisms and bind humanity together, the essential social task of a religion. Investing the engineering profession as the new lay clergy, Saint Simon found most of his disciples among the students of the Parisian Polytechnique who celebrated the new industrial gospel in famous alternative festivals in the outskirts of the French capital city. It was in Paris, through their travels of learning, or in the political economy classes at the Lisbon Polytechnic school, that Portuguese engineers became acquainted with Saint Simonianism.[37]

In 1851 engineers took political power in Portugal with the aim of applying the Saint Simonian principles to the whole of Portuguese society through a military coup that founded the Regeneration regime that dominated the politics of the country until the early 1890s.[38] With a clear parallel with the example of Napoleon III in France, the Regeneration coup, "a coup to end all coups," promised to put an end to violent political and social disputes between liberals and absolutists, and between different liberal factions that, as mentioned, characterized the first half of the century. As an alternative, Regenerationists put forward an ambitious program of "material improvements" of which railway lines constituted the most visible element.[39] Significantly enough, the leading figure of the regenerationist party—Fontes Pereira de Melo—was an engineer, one of the first to graduate from the new Army School, a member of the new generation of engineers who started their careers after the civil war. He intended to regenerate the country through the *Ministry of Public Works, Commerce, and Industry* newly created in 1852 and called

[36] Macedo, *Projectar e Construir*.

[37] *On Saint Simonian Ideology*, see Antoine Picon, *Les Saint-Simoniens*, (Paris: Belin, 2002); Pierre Musso, *Télécommunications et Philosophie des Réseaux*, (Paris: PUF, 1998).

[38] Macedo, *Projectar*, pp. 145–164. For a general overview of the political nature of the regime, see M. Villaverde Cabral, *O Desenvolvimento do Capitalismo em Portugal*, (Lisboa: Regra do Jogo, 1976); Pedro Tavares de Almeida, *A Construção do Estado Liberal. Elite Política e Burocracia na Regeneração, 1851–1890*, (Lisboa: FCSH/UNL, 1995); Maria Fátima Bonifácio, *O Século XIX Português*, (Lisboa: ICS, 2005); David Justino, *Fontismo*, (Lisboa: Edições D. Quixote, 2016).

[39] Villaverde Cabral, *O Desenvolvimento do Capitalismo em Portugal*, p. 163.

himself a "railway fanatic."[40] Indeed, an extended infrastructure of communication was one of the defining elements of Saint Simonians, dreaming with the crossing of the globe with roads, canals and railways, unifying humankind through the unrestricted circulation of goods, people, and ideas.[41]

Railways were no doubt the main political topic of the regeneration years. More than discussing if they should exist or not, deputies in the parliament embarked in passionate debates on how to fund new railway lines, how many to build, which ones to prioritize, or what should their trajectory be. After earlier failed attempts of funding a major public works company through private capitals, Fontes and his fellow engineers had it clear that the state would have to guarantee the return of railway company shareholders. Joaquim Lobo d'Ávila, another of the prominent engineers of the new generation, by occasion of a parliamentary speech on the railway connection between Lisbon and Oporto asserted that,

> Without the action of the Government we cannot take a step in the path of material progress… [Michel] Chevalier sustains the principle of a minimum interest assured by the state as the indispensable minimum to develop in a convenient scale public works companies; and says that without government impulse these companies would barely be able to emerge and progress.[42]

The Saint Simonian lesson learned at the Polytechnic and at the Army School hadn't been lost. Portugal could only take part in the new industrial order if there was enough credit to fund an extended transportation infrastructure. Considering that the Portuguese state since the independence of Brazil had drastically reduced its funding capacity, public debt would be the solution. Fontes was indeed able to find credit for Portuguese railways among London bankers. Debt would thus fund a communication infrastructure that, according to engineers, should liberate the productive capacity of the country, placing it in the "path of progress."

Although late in comparison to the rest of western Europe countries—the first line was only started in 1853—Portugal would indeed build its railway network.[43] In the initial phase of railway lines construction, the work of Portuguese engineers was not very visible, their main interventions being limited to discussions of the routes chosen by companies that added to their international capitals foreign engineers, namely of English and French origin. But after the first years, railways proved to be an excellent opportunity for Portuguese civil engineers to show their proficiency, fully participating in the design, construction, and administration of the new lines.

The main line running south/north connected the two major cities—Porto and Lisbon (1862)—while lines to the east connected the country with Spain (1863, 1880). Other lines promoted commercial agriculture by expanding Lisbon and Porto hinterlands: to the south (1864),

---

[40] J. P. Oliveira Martins, *Portugal Contemporâneo*, (Lisboa: Lello Editores, 1981 edition), p. 288. The book was first published in 1881. Maria Filomena Mónica, *Fontes Pereira de Melo*, (Lisboa: Assembleia da República, 1999).

[41] Picon, *Les Saint-Simoniens*.

[42] Lobo d'Ávila, DCD, 25.01.1854, p. 161.

[43] On the development of the railway network in Portugal see, Maria Fernanda Alegria, *A Organização dos Transportes em Portugal* (1850–1910): *As Vias e o Tráfego* (Lisboa: Centro de Estudos Geográficos, 1990).

Figure 3.5: **Progresso Repentino** (Sudden Progress). Cartoon of the minister and engineer Fontes Pereira de Melo.

The *Illustração* is an example of the illustrated press that aimed at promoting the new religion of Progress. In this image, the engineer Fontes Pereira de Melo is portrayed as a bionic creature, incorporating parts of a train, the major symbol of his political agenda of "material improvements". Fontes, the most powerful politician in the country in the second half of the nineteenth century, is surrounded by other icons of progress: telegraph, balloon, steamships, railways, and street lighting.

© Public domain. In *Illustração Luso-Brasileira*, 6 November 1858. Available online at http://hemerotecadigital.cm-lisboa.pt/OBRAS/IlustrLusoBrasil/1858/N45/N45_master/N45.pdf

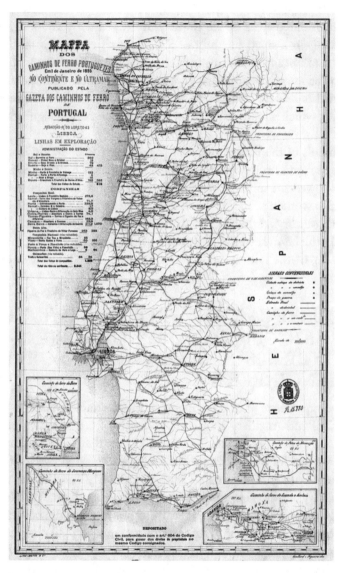

Figure 3.6: **Map of the Portuguese railroad network in 1895**. Railways, which Fontes Pereira de Melo considered embodied progress, were the corner-piece of the Regeneration period (1851-1890). At the end of the period, Portugal had over 2,000 km of railways, most of them designed under Fontes Pereira de Melo's rule.
© Public domain. In *Gazeta dos Caminhos de Ferro de Portugal*, January 1895.

new wheat production areas expanded to supply the capital city with grain; in the north, the Douro railway (1887) promoted the expansion of Port wine (the first Portuguese export) vineyards into previously isolated areas of the country further east. By 1890, 2,000 km of railways and 8,500 km of roads in a small territory of 92,000 km$^2$, connected all the district capitals of the country. While the railway didn't produce the idyllic Saint Simonian industrial utopia, it did contribute to the effective formation of a national market. The dimensions of the latter were always too small to sustain levels of industrialization equivalent to those of richer European countries, a fact which shouldn't make us ignore the historical significance of railways in Portuguese history.[44]

The main symbols of the Regenerationist period were to be found in railway stations, in the new engineering schools, or in Oporto's Crystal Palace built for the 1865 International Exhibition. Engineers sowed the territory with a new kind of monument built in an international language but that also worked as national cohesive element. The capital city guides of the second half of the 19th century started to add to the usual historical ex libris like the Jerónimos Monastery or the Bethlehem Tower, descriptions of the new big avenue inaugurated in 1880 designed by engineers, of the railway network that surrounded the city, or of the Polytechnic School.

## 3.4   CONQUERING A NEW PROFESSIONAL TERRITORY – THE ASSOCIAÇÃO DOS ENGENHEIROS CIVIS PORTUGUESES (ASSOCIATION OF PORTUGUESE CIVIL ENGINEERS)

Acknowledging the importance of civil engineers in the Portuguese state, the government created in 1864 the Civil and Auxiliary Engineering Corps of the Ministry of Public Works. It was constituted by 115 engineers, 18 architects, and 175 auxiliaries divided in 5 sections: public works, mining, water and forests, geography and telegraphy. Members of this Corps had for the first time their own career differentiated from the military one, despite the training in the Army School as military engineers of the large majority among them. Military engineers of the new Corps had to give up their position in the Ministry of War for the new one in the Ministry of Public Works, changing their identity from military to civilian. This was not consensual and resistance among those engineers that didn't want to give up their military status was strong enough to reverse the decision, leading to the elimination of the Engineering Corps in 1869, only to be finally reestablished in 1886. Engineers that insisted on the importance of a civil identity strongly manifested their discontent when the Corps was abolished. They first addressed the

---

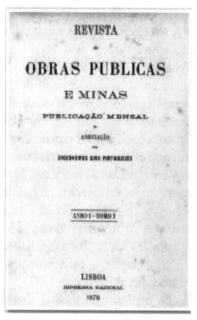

Figure 3.7: **Cover from the first number of the journal** *Revista de Obras Públicas e Minas, 1870* (*Journal of Public Works and Mines*). The journal was the mouthpiece of the Portuguese Association of Civil Engineers (1869–1937) the professional organization of Portuguese engineers. The association used the journal to advance its notion of professional identity based on expert technical knowledge and modernization of the country through infrastructures.
© Public domain.

government by presenting a petition to revoke the decision,[45] and in 1869, they moved to a more radical solution, the founding of their own professional association—the *Associação dos Engenheiros Civis Portugueses* (Association of Portuguese Civil Engineers). This Association would prove crucial in defining the self-identity of Portuguese engineers in the next decades.

The Association of Portuguese Civil Engineers had two main declared goals: First, to build a well-grounded *corpus* of engineering knowledge, in order to clearly establish the borders of this professional field, only those who held the intellectual keys for the understanding of the specific language, theoretical concepts and practical know-how of engineering—now fully considered as a science—could call themselves engineers. Second, to show to the general public how engineers played a crucial role in the production of a modern industrialized society. The Association strived to define Portuguese engineers as a specialized group within society, possessing

---

[45] *Representação dirigida aos Senhores Deputados da Nação Portugueza pelos Engenheiros e Conductores do Extincto Corpo d'Engenharia Civil e seus Auxiliares (Petition Addressed to the Members of the Portuguese Parliament by the Engineers of the late Corps of Civil Engineersand Assistant Engineers)*, (Lisboa, 1869).

and deploying technical knowledge essential to the modernization of the country; its promoters sought, therefore, the creation of a well-defined professional consciousness as well as the public recognition of engineers as a social-cultural fact.

The first President of the Portuguese Association of Civil Engineers, founded by 107 male members[46] (the number of members continued to grow throughout its 68 years of activity) was João Crisóstomo de Abreu e Sousa, an engineer who embodied the changes in the engineering profession of the 19th century: trained in the old Academy of Fortification and later in the Parisian École des Ponts et Chaussées; active participant in the civil war on the liberal side; involved in the design and construction of roads, canals, and later of railways; member of the new bureaucratic elite of the Regenerationist period as member of the Council of Public Works and Director of Public Works; promoter of reforms in technical training in the Army and Polytechnic Schools; prominent politician, member of parliament, Minister of Public Works, Minister of the Navy and Overseas, and Minister of War.[47] The choice of João Crisóstomo as leader of the Association, in addition to recognizing his many professional merits, thus consecrated the strong ties between technical proficiency, transportation infrastructure, state bureaucracy, and politics all weaved together by the engineer's biography.

The bylaws of the Association of Portuguese Civil Engineers, written and approved in 1869, established the official architecture of the professional group, reflecting not only the principal areas of engineering in Portugal—railways, ports, roads, and mines—but also those that were still marginal and aspirational, such as industry. Civil engineers broke apart from their military counterparts in a conscious and planned strategic move. It was a powerful professional statement, almost a leap of faith of all these engineers who were by training, one should keep in mind, military engineers. This strategy paid off well as demonstrated by the fact that by the

---

[46] Portuguese engineering is a male world until very late. The first Portuguese female engineer was Rita de Moraes Sarmento, born in 1872. When she was 15 years old, she enrolled at the Polytechnic Academy of Oporto and concluded the 2-year Public Works course in 1894. In 1896, she received the certificate for the pursuit of the profession, but she never worked as an engineer. She married in 1898 to António dos Santos Lucas, Ph.D. in Mathematics and Lieutenant of Engineering and dedicated her life to support her husband's long and successful professional career. The second female-engineer was Maria Amélia Ferreira Chaves, who enrolled at the Instituto Superior Técnico (Technical Institute, Lisbon) in 1931. She graduated as a civil engineer in 1937, almost half a century (43 years) after Rita Sarmento. Unlike her, however, Maria Amélia worked as an engineer in the service of the Lisbon City Council. She was able to engage in fieldwork, supervising construction sites, signing projects and monitoring their execution. She even invented a special pant skirt to be able to walk on the scaffold without embarrassment. Feeling that her work was minimized she presented her resignation in the 1940s. As an independent or on behalf of others employees she continue to work as a civil until she was ninety years of age. She was the first woman to register in the Ordem dos Engenheiros (the Portuguese professional association that rules the pursuit of the profession) in 1938. She was also the author of the first anti-seismic test carried out in Portugal. In the late 1940s, four more women graduated as civil engineers: Maria da Conceição Moura in 1947, Virgínia Faria de Moura and Maria Emília Campos Matos in 1948, all at the Faculdade de Engenharia (Faculty of Engineering of the University of Porto), and Maria da Conceição da Costa Barros Magalhães Cruz Azevedo, in 1949 (Technical Institute). All of them worked as engineers and registered in the Ordem dos Engenheiros. See Júlia Coutinho, *Estudos Sobre o Comunismo. Movimentos Radicais da Esquerda e a Oposição ao Estado Novo.* https://estudossobrecomunismo.wordpress.com/2012/06/03/julia-coutinho-mulheres-pioneiras-em-engenharia-civil.

[47] Maria Paula Diogo, "João Crisóstomo de Abreu e Sousa," *Biografias de Cientista e Engenheiros Portugueses online,* (Biographies of Portuguese Scientists and Engineers online) http://www.ciuhct.com/index.php/pt/biografias.html.

Figure 3.8: **Portrait of João Crisóstomo de Abreu e Sousa by Veloso Salgado, 1890**. João Crisóstomo de Abreu e Sousa was the first President of the Portuguese Association of Civil Engineers. He embodied the trajectory of nineteenth century Portuguese engineers: trained in the old Academy of Fortification and later in the Parisian École des Ponts et Chaussées; active participant in the civil war on the liberal side; involved in the design and construction of roads, canals, and later of railways; member of the new bureaucratic elite of the Regenerationist period as member of the Council of Public Works and Director of Public Works; promoter of reforms in technical training in the Army and Polytechnic Schools; prominent politician, member of parliament, Minister of Public Works, Minister of the Navy and Overseas, and Minister of War.
© Public domain. The portrait is at the D. Maria II Room in the Portuguese Parliament building.

end of the century the number of engineers who had a specific training in the civil area would match those with military training.

The job market for engineers also makes explicit the very specific conditions that led to the creation of a community of non-military engineers: as railways were the main justification for expanding the making of engineers in Portugal, the typical 19th-century engineer was a civil servant, working either in public works (railways, ports, bridges) or directly in government (as ministers or members of various committees).[48]

Despite the lack of regular dialogue with industrialists, engineers continued to envisage industry as a potential prime area of intervention, as the multiple texts and field initiatives prove (see Chapter 4). In the last decade of the 19th century, although public works continued to

---

[48] Ana Cardoso de Matos and Maria Paula Diogo, "Le role des ingénieurs dans l'administration portuguaise: 1852–1900," *Quaderns d'Història de l'Enginyeria*, X, (2009), pp. 351–365.

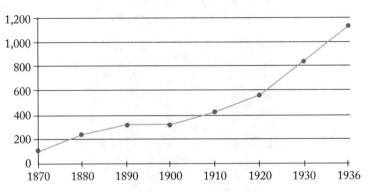

Figure 3.9: **AECP** members. Source: *Revista de Obras Públicas e Minas* (*Journal of Public Works and Mines*) (1870–1936).

occupy the most significant slice of Portuguese engineering, these would gradually make space for a diversification of interests centered on industrial activity. A new image of the engineer as both the efficient technical and the social leader, which, as we shall see, characterized the 20th century was already emerging

The areas related to public works—mainly railways, ports and bridges and, later, the electrical system—were, not surprisingly, the main subject of the articles published in the mouthpiece of the Association, the *Journal of Public Works and Mines*. Industrial topics remained scarce confirming, with some few rare exceptions, the gap and fragile dialogue between engineers and the industrial milieu. Until the early 20th century, few Portuguese engineers could expect to apply and develop their expertise outside the area of public works structures. Consequently, the academic training of an engineer became essential as a prerequisite to enter the hierarchical structures of the civil service.

Portuguese civil engineers fashioned themselves as an elite bound to occupy dominant state positions, a true *noblesse d' état* as already mentioned before.[49] They managed to achieve this status both by training and professional strategy, as well as by their strong relationship with the state.[50] The formal transmission of technical knowledge and skills in a restricted set of higher education technical schools, the value of the diploma as a piece of legitimation of authority, the building of a professional society that secured the *esprit de corps*, and the homogeneity of its members, all contributed to produce an elite aiming at occupying dominant positions. The transmission of knowledge and practices of technical competence was part of a process of transmission of social and political power among individuals with the same academic and professional

[49] Bourdieu, *La Noblesse d'État*.
[50] Cardoso de Matos, Diogo, "Le role des ingénieurs."

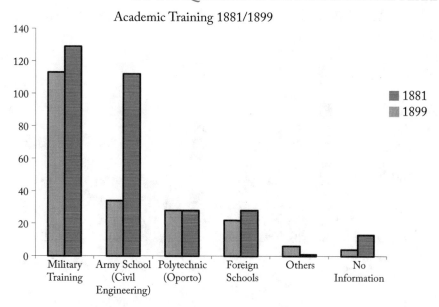

Figure 3.10: Academic training. Source: *Revista de Obras Públicas e Minas* (*Journal of Public Works and Mines*), 1881, XII, pp. 135–136 (March/April), pp. 45–58; 1899, XXX, pp. 351–352 (March/April), pp. 167–201.

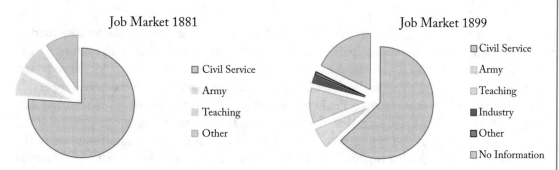

Figure 3.11: Job market. Source: *Revista de Obras Públicas e Minas* (*Journal of Public Works and Mines*), 1881, XII, pp. 135–136 (March/April), pp. 45–58; 1899, XXX, pp. 351–352 (March/April), pp. 167–201.

qualities, very close, as already mentioned, to the consecration of a title of nobility.[51] Moreover, the public image of the engineer is merged with the national agenda of modernization and the gospel of progress, endowing the engineer with a messianic dimension.

[51] Maria Paula Diogo, "Portuguese engineers, public works, and professional identity. The Portuguese Association of Civil Engineers (1869–1937)." *HoST* 7, (Spring 2013)/Special Issue: pp. 67–84.

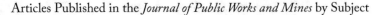

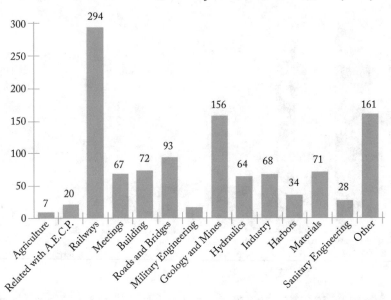

Figure 3.12: Articles published in the *Journal of Public Works and Mines* by subject. Source: *Revista de Obras Públicas e Minas*.

Portuguese civil engineers asserted their professional essence through a very active participation in building the liberal state, particularly during the second half of the 19th century.[52] They were the anchors of the *Regeneration* and embodied, as a professional group, its ideology and its project for Portugal, close to the Saint Simonian ideal of governance by technical experts.[53] The engineers of the Association claimed for themselves the role of leaders of modernity; they were, as they sated in their journal at the end of the 19th century, "the avant-garde and the guardian[s] of progress and civilization."[54] They promised the technical expertise to build the new technical landscape that would allegedly place Portugal once more in the arena of developed European countries, after it had lost most of its empire in the early 19th century. By turning

[52] Luís Miguel Carolino, Teresa Salomé Mota, and Dulce Figueiredo, "The Polytechnicians of the Portuguese long 19th century: technical expertise, military aspirations, and political disenchantment. A preliminary study," *HoST* 7, (2013); Cardoso de Matos and Diogo, "Le role des ingénieurs;" Diogo "In search of a professional identity."

[53] This was also not far from William Henry Smyth's concept of ruling the people through the agency of engineers and scientists William Henry Smyth was a Californian engineer who coined the word "technocracy" to describe "the rule of the people made effective through the agency of their servants, the scientists and engineers." Although the word had been used before its meaning stabilized Smyth's 1919 article "'Technocracy'—ways and means to gain industrial democracy," in *The Journal Industrial Management*, (57).

[54] "Relatorio e contas da associação dos engenheiros civis Portugueses com referencia ao ano de 1897," *R.O.P.M.*, vol. XXIX, 337/339 (1898), p. 4.

their own professional narrative into a national narrative, they performed as no less than organic intellectuals, producers of liberal cultural hegemony.

CHAPTER 4

# Engineers, Industrial Workers, and the Bourgeois City

## 4.1 SAINT-SIMON IN PORTUGAL AND THE FOUNDING OF THE LISBON INDUSTRIAL INSTITUTE

The engineer J. Ribeiro de Sá was one of the organic intellectuals of the period. He had been trained at the Lisbon Polytechnic School and had an important presence in the liberal public sphere as the main editor of one of the most ambitious tittles of the Lisbon press, the *Revista Universal Lisbonense*. In the Fall of 1849, when covering the opening of the Exhibition of Portuguese Industry organized in the facilities of the Navy Arsenal, he didn't restrain his enthusiasm and emphatically asserted that the exhibition was a "national celebration, a reunion of brothers, even when it is the work and not the blood that binds the thoughts and the arms of hundreds of men."[1]

For those familiar with the political jargon of the era, this use of terms in reference to industry is readily identifiable with the afore mentioned theories of Henri de Saint-Simon and its technological utopian socialism. Ribeiro de Sá, also held no doubts concerning the cohesive effects of technology in the social body. When describing the display of the José Pedro Collares factory at the exhibition, one of the main foundries in the city, Ribeiro de Sá made metalworkers the embodiment of progress through the transformative power of technology: "the dark and robust arms moved by the breeding of steam" converted the 60 foundry workers into the "glorious race of the most useful men to society."[2] The Industrial Exhibition was the alleged proof that there was no hatred between workers and owners, all united through steam power in a new society of industrialists. According to Saint Simonians, steam engines didn't produce the society

---

[1] See Sociedade Promotora da Indústria Nacional, *Exposição da Industria em 1849*. For a more detailed description of the exhibition, see Saraiva, *Ciencia y Ciudad*, pp. 167–169. Although preceded by few other similar events, it seems reasonable to perceive it as the first big scale experiment in Lisbon with industrial exhibitions, a crucial element of 19th-century urban life, epitomized by the Great Exhibition two years later in London. The same industrial association—the Society for Promoting National Industry—that put together the Portuguese industry exhibition also undertook responsibility for showing two years later the Portuguese products at the London Crystal Palace. Ribeiro de Sá would also be the one in charge of the Portuguese representation in London in 1851. On this association see, Cardoso de Matos, *Ciência, Tecnologia e Desenvolvimento Industrial*, pp. 138–142 and pp. 266–67; S. J. Ribeiro de Sá, "Sociedade Promotora da Indústria Nacional." An extended version of this chapter was published in Tiago Saraiva and Ana Cardoso De Matos. "Technological Nocturne: The Lisbon Industrial Institute and Romantic Engineering (1849–1888)." *Technology and Culture*, 58.2 (2017): 422–45.

[2] Ribeiro de Sá, "Sociedade Promotora."

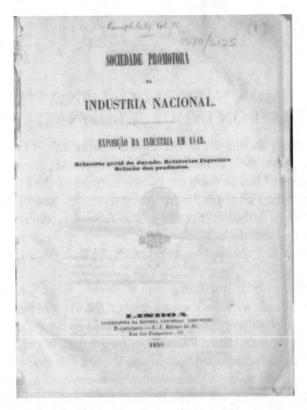

Figure 4.1: **Report of the jury of the Portuguese Industrial Exhibition of 1849 published by the** *Sociedade Promotora da Indústria Nacional*.

When covering the opening of the Exhibition, the engineer Ribeiro de Sá described it as a "national celebration, a reunion of brothers, even when it is the work and not the blood that binds the thoughts and the arms of hundreds of men." Sá was mobilizing Saint Simonian rhetoric that made technology the binding element of society.

© Public domain.

of individuals celebrated by liberal thinkers or the class society later denounced by Marxists, they formed instead a new human brotherhood.

Engineers as main promoters of Saint Simonian ideology reserved an important place for themselves in this new technological brotherhood. As reminded by Joao Crisóstemo when assuming the role of president of the association of civil engineers[3]:

> Engineers play an important role as the main intermediates between both the entrepreneur and the worker and the school and the factory(…). A country where engineers do not perform such role is doomed to be underdeveloped and always to depend on foreign countries.

No less important than communications in the political economy catechism of Saint Simonians explored in the previous chapter was the issue of industry and machines as binding elements of a society of producers. Although the new Ministry of Public Works created in 1852 after the regenerationist coup also included "Industry" in its denomination, thus bringing explicitly to the structure of the Portuguese state the trilogy—Communications, Commerce, and Industry—industry has played a secondary role in historical accounts of the regime, namely when compared to railways. After all, one has to keep in mind that in 1852 the steam engines used in the whole of the Portuguese industry didn't produce more than 1,000 hp, while in Prussia that number reached some 42,000 hp.[4] In 1851, Portugal had no more than some 16,000 industrial workers, but from 1852–1890 the number of Lisbon industrial workers increased more than three times. Also, by 1881 an industrial enquiry counted already some 10,000 hp powering the capital's factories.[5]

Portugal did industrialize during the second half of the 19th century, propelled by protectionist tariffs, but it did so considerably slower than the rest of Western Europe, its industrial growth always limited by the small dimension of its internal market. It is not our intention to add new arguments to the causes of Portuguese backwardness in respect to European countries. And one should keep in mind that such metrics of progress were themselves first put in place by engineers who were the ones who organized industrial enquiries and compared countries to one another through the machines presented in industrial and world exhibitions. We would like to suggest instead that it is rewarding to take seriously the Saint Simonian commitment to industry of the engineers dominating the regenerationist regime. Indeed, while hp numbers placed the

---

[3] "Progressos das sciencias e das artes com relação à engenharia civil nos fins do século XVIII. Discurso pronunciado pelo ministro e secretário d'estado honorário João Crysostomo de Abreu e Sousa, no acto de tomar posse da presidência da associação dos engenheiros civis, em 20 de janeiro de 1872," *R.O.P.M*, III, no. 26, 1872, p. 59.

[4] Portuguese economic historians have demonstrated that Regeneration oversaw Portuguese industry growing on a sustained trend, propelled mostly by expanding internal demand, a dynamics concentrated first and foremost in the capital city. The numbers are not consensual but they all point at the same general trend. Jaime Reis, "O atraso económico português em perspectiva histórica"; Pedro Lains, *Os Progressos do Atraso*; David Justino, *A Formação do Espaço Económico Português*.

[5] Conceição Andrade Martins, "Trabalho e condições de vida em Portugal, 1850–1913."

country at the tale of industrializing Western Europe, no other nation could boast of a ruling elite of engineers as committed to the Saint Simonian gospel as the Portuguese one.[6]

This engagement with Saint Simonian ideals, namely the emphasis on industry and technical education, is again crucial to understand the founding of the *Instituto Industrial* (Industrial Institute) in December 1852.[7] This Institute was renamed in 1911 as Technical Superior Institute (*Instituto Superior Técnico*), today still the main engineering school in Portugal.[8] Its first director was José Vitorino Damásio, one of the main characters responsible for the success of the military operations of the regenerationist coup of 1851. Trained in Mathematics by the University of Coimbra, he had proven his valor in battle in the civil war opposing absolutists and liberals and in the following confrontations between different liberal factions, always on the side of the most progressive and revolutionary parties. His mix of political enthusiasm, physical courage, and technical training was not lost in organizing sieges, commanding military columns or planning defense strategies.[9]

When regenerationists came to power it is thus no surprise to find Damásio as member of the Council of Public Works, a restricted group of six engineers in the Ministry who had to give their seal of approval for the building of any new major communication infrastructure in the country.[10] If the other members of the powerful council bragged about their expertise in civil engineering, Vitorino Damásio was clearly the one closest to the industrial world. He was an accomplished industrial entrepreneur, co-owner of a successful foundry in Oporto, the second city of the country, and deeply involved with other members of the local elite in projects of social reform. Inspired on Saint Simonian ideals of association and in the role of technical education for workers' emancipation he founded the Industrial Association of that city in which he became the main promoter of an Industrial school with the aim of improving industry by the application of science. It was this Oporto experience, with its emphasis on "observation, manipulation and experiment, as well as the graphic works" as opposed to mere "oral lectures," that would serve

---

[6] Probably, only in Brazil can one identify a similar historical example of political power take over by Saint Simonian engineers. On an exemplary Brazilian Saint Simonian, see Susanna B. Hecht, *The Scramble for the Amazon and the Lost Paradise of Euclides da Cunha*, (Chicago: Chicago University Press, 2013).

[7] Together with the also newly founded Agricultural Institute, the Industrial Institute was put under direct dependence of the Ministry of Public Works, Commerce, and Industry, in clear contrast to the other higher education institutions of the country, leaving no doubts about its political patronage. Significantly, the new Ministry controlled neither the Polytechnic School nor the Army School, the two main engineering institutions at the time.

[8] Although Portuguese historians of science and technology have put big emphasis on engineering institutions such as the Polytechnic School, the Industrial Institute has received considerably less attention. The best-informed source, although focused on the particular case of chemistry, is the very recent dissertation by Isabel Cruz, "Da Prática de Química à Química Prática." See also, from an institutional history of education point of view, Mário A. N. Costa, *O Ensino Industrial em Portugal de 1852 a 1900*; Luís Alberto M. Alves, "Contributos para o estudo do ensino industrial em Portugal." Joaquim Ferreira Gomes, *Escolas Industriais e Comerciais do Século XIX*, Coimbra, 1978, pp. 85–151. For later periods of the Superior Technical Institute see the commemorative volume, Jorge Freitas Branco Ed., *Visões do Técnico no Centenário*.

[9] For biographic details on Vitorino Damásio see, J. F. Nery Delgado, "Elogio histórico de José Vitorino Damásio;" Jorge Fernandes Alves and José Luís Vilela, *José Vitorino Damásio e a Telegrafia Eléctrica em Portugal*.

[10] On the Council of Public Works (Conselho de Obras Públicas) see Marta Macedo, *Projectar e Construir a Nação*, pp. 165–187.

Figure 4.2: **Portrait of José Vitorino Damásio by Francisco José de Resende Vasconcelos, 1853**.
Professor at the Porto Polytechnic Academy and industrial entrepreneur in the same city, José Vitorino Damásio was also the first director of the Lisbon Industrial Institute founded in 1852. He combined different elements of romantic engineering: military heroism, enthusiasm for steam engines, and commitment to technical education as workers' emancipation.
© Public domain.

as first source of inspiration for the new Lisbon Industrial Institute of which Vitorino Damásio assumed the direction.[11]

## 4.2   WORKER'S EMANCIPATION THROUGH TECHNICAL EDUCATION

In Vitorino Damásio's report on the first year of the Industrial Institute, he mentions together with the Oporto school some other former initiatives, mostly in Lisbon, directed at training "the industrial class," a typical saint-simonian phrase he used to refer to all the social strata involved in industrial activities, from workers to engineers, from foremen to factory owners.[12] When explaining the previous failures, Damásio first points out the issue of the target student population public. Past institutional arrangements had arguably not been inclusive enough, aiming only at specialized students with previous formal training. The new Institute, in contrast, was to offer three kinds of courses—elementary, secondary, and complementary—and, to be admitted, candidates had only to be 12 years old and know how to read, write, and count.[13]

This three-level structure of the courses, aimed at training skilled craftsmen and foremen, but also industry directors, made the Lisbon Industrial Institute a combination of the French model of the *Conservatoire des Arts et Métiers* geared toward workers' education and the elitist *École Centrale* geared toward engineers.[14] In Lisbon, all different levels were brought under the same institutional roof. The traditional hierarchy of the workshop divided in worker, official, and foreman (mestre) was thus complemented with a new one—the director (mechanic or chemical). Those willing to achieve this new higher status of director had to attend classes in applied chemistry, industrial mechanics (taught by Damásio), as well as in economy and industrial legislation. The chair of the latter course was no other than Oliveira Marreca, the first expert in the country on the political economy of the French Michel Chevalier, the most accomplished of the Saint Simonian apostles.[15] And although the mechanic and chemical directors had a distinguished status in the factory floor, it is important to retain that they didn't carry the formal tittle of engineer. Despite the commitment to industry of prominent engineers such as Damásio or Crisóstomo, only in 1903 would the students graduated from the Industrial Institute be granted the tittle of industrial engineers.

[11] *Jornal da Associação Industrial Portuense*, 6, 1852, p. 86.

[12] "Abertura das aulas no Instituto Industrial de Lisboa. Relatório do Director interino do dito instituto, José Vitorino Damásio," *Boletim do Ministério das Obras Públicas, Comercio e Industria* (1854), 10, pp. 248–256.

[13] Abertura das aulas.

[14] The focus on industrial workers and their overwhelming presence in the first decades of the Institute might suggest that we are overstretching in relating it to the elitist Parisian *École Centrale*. Nevertheless, the training in Portugal of industrial engineers, the historical figure first associated with the *École*, was due mostly to the Lisbon Industrial Institute. The conversion in 1911 of the Lisbon Industrial Institute into Technical Superior Institute as the main engineering school of the country demonstrates that the initial intentions of higher education were not just rhetoric, justifying the comparison with the *École Centrale*. On the *École Centrale*, see John Hubbel Weiss, *The Making of Technological Man. On the Conservatoire*, see Robert Fox, "Education for a new age. The conservatoire des arts et métiers, 1815–1830."

[15] António de Oliveira Marreca, "Considerações sobre o curso de economia política, publicado em paris em 1842 pelo Sr. Miguel Chevalier," 1843.

The new society of industrialists produced at the Institute and made of workers, foremen, and directors, all united through technology, could only be materialized at night. As stated by Damásio, previous institutions only had "daytime lectures, which forced those workers interested in attending them to give up of part of their already meager salary."[16] For workers to join in, Industrial Institute courses only started at dusk. The Institute can thus be seen also as a device to transform the night of the workers into an edifying experience. This was another typical Saint Simonian solution, willing to overcome class conflicts by forging a coherent "industrial class" and promising that through proper training anyone could claim the heroic status of industrialist.[17] The fact is in the first year no less than 744 students attended the institute, a number that would reach 1,079 two years later.[18] Such high level of attendance is even more striking when bearing in mind that a 1852 survey indicated that the country had some 16,500 industrial workers.[19] Damásio had no doubts that such large turnout was due in great measure to all the courses being lectured after 8 pm. And it is thus no surprise to find out that one of the major items of the Institute's budget in those first years were expenses with gas illumination.[20]

As suggested by Jacques Rancière, in his as much inspired as provocative "The Night of the Proletarians," nighttime should be perceived as the only space/time left for workers' to become proper humans able to experience life in a meaningful way.[21] He makes a convincing reading of French workers' press showing their discontent with initiatives aimed at building a proletarian culture based on exulting the moral qualities of repetitive and tedious activities, and the desire of workers to have access instead to bourgeois high culture by writing poetry or playing the piano. Our case seems to indicate that together with fine arts and letters, technology was also able to embody the promise of humanizing workers. The hundreds of students that attended the Institute's classes under the gas light, the large majority (more than 3/4) of them industry workers and their sons, had allegedly found an alternative to the extended shifts in Lisbon factories as well as to the city's more than 300 taverns.[22]

Together with the classrooms and laboratories for lectures in mathematics, physics, chemistry, descriptive geometry, mechanics, and drawing, the Institute premises also included improvised additions housing model workshops as well as an industrial museum.[23] In the workshops, the scientific principles learned in the classroom were to be applied to practical tasks. The Insti-

---

[16]Marreca, *Considerações sobre o curso.*

[17] Interestingly enough the foundry owners that offered worse working conditions such as José Pedro Collares had started with small workshops and had the same social background than their workers.

[18] José Silvestre Ribeiro, *História dos Estabelecimentos Científicos*, pp. 346–358.

[19] Conceição Andrade Martins, "Trabalho e condições de vida."

[20] Damásio, "Abertura das aulas no Instituto Industrial," pp. 248–256.

[21] Jacques Rancière, *La Nuit des Prolétaires.*

[22] Ribeiro, *História dos Estabelecimentos Científicos*, pp. 346–358.

[23] The above mentioned Ribeiro de Sá, in his condition as Portuguese commissioner to the Great Exhibition in London in 1851, seized the occasion to make a technological Grand Tour of Europe to purchase collections for the Institute's museum, confirming the tight connections between the Institute and Industrial exhibitions. In 1855, it was Damásio himself who profiting from his visit to the Paris Universal Exhibition brought in new machines and instruments. *Diário da Câmara dos Deputados*, 30/6/1852, pp. 55–56. For the importance of these travels of learning for Portugues engineers see Maria Paula Diogo and Ana C. Matos, *Travels of Learning.*

tute had from its onset workshops of carpentry, foundry, forging, sawing, drawing, and precision instruments. Except for the last one, which constituted a total novelty in the country, all of them were conceived as model versions of the real industries surrounding the Institute.

For Damásio, the "workshops were to be not only a practical class but also a factory, producing enough to fund itself."[24] The Institute was a true factory, although an exemplary one, not only for its level of technical sophistication but as well as for its social and working conditions. The common humiliating atmosphere of the factory floor was to be abolished with no tolerance for any kind of physical punishment. In contrast to the low level of formality of the contracts dominating Lisbon industrial world, in the Institute's workshops a set of rules made more transparent the nature of the work, its duration, as well as the punishments for noncompliance (salary cuts, expulsion). If the Institute contributed to standardize working processes, it also meant to be exemplary in demonstrating the opportunities for social mobility thus opened. Those students singled out for the quality of their work could become decurions—responsible for teaching ten other students—and later directors of a workshop themselves. The final aim was, always following Damásio, that after getting enough theoretical and practical training they would open their own independent business.

As good proof of the actual success of Damásio's proposals, five years after its opening, the Lisbon foundries' owners started to pressure the government for the closing of the Institute's workshops arguing that these constituted unfair competition to their own businesses, attracting namely an increasing number of workers in a context of scarcity of workforce. In 1858 an official committee was formed to inquire "if the practical lessons of the Industrial Institute workshops were detrimental to private industry," which led to the closing of the workshops in 1860, with the important exception of the precision instruments one.[25] While it is common for historians of technology to insist on the role of technical education as an extension of social control over workers, such initiative by Lisbon factory owners should make us think as well of the emancipatory value of such education.

## 4.3   THE INDUSTRIAL INSTITUTE AND CITY REFORM

In order to understand the full historical significance of this Industrial Institute, which would evolve as we will see into the main engineering school of the country, one can't just look at its interior, regulations and classes. Its location in the city and relation to its surroundings is also important. Actually, and according to those Lisbon factory owners that led to the close down of the Institute's workshops, the problem was it had fused too well with its industrial surroundings. Both the location and architecture of its facilities contributed to the confusion. The Institute occupied the old building of the *Paço da Madeira* (wood customs) to which it were added several improvised barracks for workshops, a typical arrangement of the liberal age of taking over spaces

---

[24] Damásio, "Abertura das aulas."

[25] João Palha de Faria Lacerda, João Cordeiro Manuel, and José Ennes, "Relatório da comissão de inquérito nomeada por portaria de 21 de junho de 1858," p. 713.

Figure 4.3: **The Industrial neighborhood of Boavista, 1906.**
The Lisbon Industrial Institute (bottom right) was located in the most industrialized area of the city, the Boavista, next to gasworks (bottom left), the mint, the navy arsenal, and several foundries.
© Arquivo Fotográfico da Câmara Municipal de Lisboa.

previously used by Old Regime institutions.[26] More significantly, it was located in the Boavista neighborhood, the major industrial area of the city. In fact, a look at the map reveals its proximity not only to several foundries (Bachelay, Vulcano, Phénix, José Pedro Collares), but also to the mint (equipped with steam engines since 1835), the navy arsenal, the gasworks, as well as smaller units producing glass, ceramics, beer, and ice.[27] The Industrial Institute thus contributed to make the Boavista the most densely packed industrial area of Lisbon.

The first justification for such industrial concentration in the area was its closeness to the river Tagus. The gasworks, located in the Boavista since 1847, offers the best example of the advantages associated with the river from where it got its main raw material—coal imported

[26] Most of the institutions of the new liberal regime in Portugal, as in many other European countries, occupied palaces that had belonged to the aristocracy or convents and monasteries of the religious orders. For a systematic overview of the Old regime buildings taken over by liberal scientific and technical institutions see, Saraiva, *Ciencia y Ciudad*.

[27] Jorge Custódio, "Reflexos da industrialização na fisionomia e vida da cidade. O mundo Industrial na Lisboa Oitocentista," *O Livro de Lisboa*, pp. 435–492.

from England. As many other companies in the neighborhood, the gasworks had its own private dock granting direct access to the river. Also, the prevailing north western winds directed most of the fumes of its highly polluting operation toward the Tagus, in a context in which increasingly tight regulations forbade the presence of the most contaminating industries inside the city perimeter.[28] And if transportation and environmental considerations were common to all other industries in the Boavista, the low quotas of the riverbank enabled as well the gas company to supply the entire city through its distribution network dispensing with big pressures in its pipes.

It is common to underline in studies of geography of innovation the advantages derived from the presence of different industries in the same area.[29] In the case of the Boavista, while the local foundries supplied tubes for the expanding gas infrastructure, they got to be the first industrial establishments to benefit from gas lighting, which enabled them to extend their working hours. We've mentioned also how much the Industrial Institute in its purpose of illuminating the nights of the workers depended on the gasworks next door. In the opposite direction, the gas works also took the Institute as source of technical expertise and instruments of precision. The most telling case is the one of the Institute's student Emílio Dias who, after excelling in both chemical analysis and building precision instruments at the Institute, was hired by the Gas Company in 1872.[30] In his new job Dias would patent an electric manometer that controlled pressure in the retorts used for distilling operations, leading to important savings in coal consumption, the main cost of the company.[31] The presence of the engineering school in the Boavista thus constituted a new important factor for the success of local industries, another important variable in addition to the easy access to raw materials or to the products from other industries in the area.

For all its economic advantages, the Boavista was an infamous area. Adding to industrial pollution, the concentration of workers and prostitutes made it a place to avoid by respectable bourgeois city dwellers. The surge in Lisbon of violent cholera and yellow fever epidemics in the years of 1856 and 1857 only made problems more obvious. After some 9,000 deaths, the entire scientific community of the capital was mobilized to discuss and propose urgent measures to sanitize the city.[32] Maps produced by hygienists detailing the progression of the epidemics through the city singled out the Boavista as one of the problematic areas. The dark dense industrial area, combining factories, narrow streets, stinky docks, and taverns, was the perfect target for hygienists denouncing the nuisances associated with industrialization and the concentration

---

[28] Ana Cardoso de Matos, *A Indústria do Gás em Lisboa. Uma Área de Confluência de Várias Abordagens Temáticas*, Interestingly enough the professors of the Industrial Institute Vitorino Damásio and José Máximo de Oliveira Pimentel were charged with the task of surveying the conditions of operation of the factory of "Purgueira" oil in Alcântara due to the complaints from the neighbors.

[29] Manuel Castells and Peter Hall, *Technopoles of the World: the Making of 21st Century Industrial Complexes*, (London: Routledge, 1994).

[30] "O ensino profissional no albergue nocturno," *O Occidente*, 21/9/1888.

[31] Eduardo Pereira, another student from the Industrial Institute, was also hired by the gas company as second engineer.

[32] Tiago Saraiva, *Ciencia y Ciudad*, pp. 58–59.

Entrada da nova rua «Vinte e Quatro de Julho», junto da egreja de Santos, no aterro da Boavista

Figure 4.4: **Boavista Embankment, 1863.**
Engraving published in *Archivo Pittoresco*, the main illustrated journal in Lisbon, depicting the embankment designed by Vitorino Damásio as new place for bourgeois sociability. Note the industrial stacks that dot the area where the Industrial institute was located.
© Public Domain.

of poor populations. Júlio Máximo de Oliveira Pimentel, chemistry professor at the Industrial Institute as well as at the Polytechnic school, condemned the "monstrous edifices that do not honor neither the taste nor the intelligence of our industrialists. True industry doesn't demand neither accepts the disorder of constructions and narrow filthy streets. It doesn't ask for gray buildings darkened by the smoke of their stacks or vicious air from the unsanitary emanations of badly calibrated machines."[33]

One of the first measures to be taken to prevent future epidemics was to drain the sludge and sands that blocked the city sewage system and caused periodical inverted flow of wastewaters into the city during high tide. Vitorino Damásio, the director of the Industrial Institute was the one selected as leader of the commission formed to indicate the "preventive and hygienic ways to keep Lisbon safe from epidemics." He seized the occasion to propose the building of an embankment streamlining the river margins of the entire Boavista area. Shortly after, Damásio himself was put in charge of the project and construction work.

---

[33] Quoted in Maria Helena Lisboa, *Os Engenheiros em Lisboa. Urbanismo e Arquitectura, 1850–1930*, (Lisbon: Livros Horizonte, 2002), p. 115.

The Embankment (Aterro) was conceived to serve four different purposes: to clean the margins from noxious sludge; to allow tall ships to dock; to start a railway line along the river toward the west; and to embellish the area through the opening of a boulevard lined with trees and gaslights.[34] Hygiene, communications, and leisure, all combined in a single infrastructure that conquered some 50 m of land into the river. The new embankment avenue offered a straight axis along which to organize the industry of the area. What had been previously a mess of warehouses, small docks, and abandoned boats, had given place in 1863 to an elegant curvilinear illuminated avenue 1,160 m long and 20 m wide.

The *Archivo Pittoresco*, the main illustrated journal of the capital, offered its readers the confirmation that the city was progressing in the right direction: an engraving published that same year showed bourgeois couples strolling along the Embankment. Starting their promenade in the church in the foreground, they advanced following the tramway tracks and the line of gas lamps in the direction of two smokestacks and newly built houses.[35] The "muds of the Boavista" had been replaced by the enlightening experience of a technological march of progress. The Industrial Institute not only aimed at becoming a technical and social model for the industries surrounding it, it also contributed to make the Boavista embankment the favorite place for bourgeois sociability in Lisbon in the 1860s and 1870s.

## 4.4   OPERA, PRECISION, AND THE INDUSTRIAL INSTITUTE

While the Industrial institute promised to emancipate workers, its connections with Lisbon's bourgeois sociability in the second half of the 19th century were also profound. To demonstrate this there is no more eloquent case than referring to opera. The bourgeois readers of the illustrated journal *O Occidente*, the successor of the *Archivo Pittoresco*, were in fact used to the juxtaposition of cultural themes and technology. The same issue that reproduced an engraving of the above mentioned Industrial Institute alumni Emilio Dias with his electromechanical instrument for regulating physical and chemical purification operations in the Lisbon Gas Company furnaces, included as well the critique of the performance at the São Carlos Theater, Lisbon opera house, of Arrigo Boito's *Mephistopheles*.[36] The bringing together of the devil and lighting technology was not just a curious coincidence. The actual critique of the opera was about the technology used to produce Mephistopheles. The critic scorned the production for its "bad placing of the lights" and for not understanding that the Witches' Sabbath of the second act

---

[34] Saraiva, *Ciencia y Ciudad*, pp. 94–95.

[35] *Arquivo Pitoresco*, 1863, no. 40, p. 313.

[36] "O scenario da opera mephistopheles," *O Occidente*, March 21, 1881; Francisco Fonseca Benevides, "Manómetro electrico por Emilio Dias." *O Occidente*, March 21, 1881. Only to help our argument, the illustration of the new scientific apparatus shared space in the same page with an illustration of Camões commemorative medals.

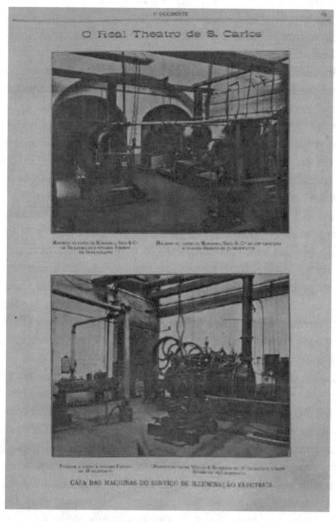

Figure 4.5: **Steam engines that powered the dynamos producing electricity for the light effects of the São Carlos Opera Theater.**
The machinery for the Lisbon opera theater was installed under the supervision of Francisco Fonseca Benevides, opera critic and professor of Physics at the Lisbon Industrial Institute, and later its director. Engineering, as understood by Benevides, was a romantic practice in which opera, thermodynamics and industry were all intertwined.
@ Public Domain. In *O Occidente*, 1902.

demanded more than well-crafted scenery."[37] The review was not signed, but chances are high Francisco Fonseca Benevides wrote it. Not only was Benevides a regular contributor to the journal (he also wrote the note on Emilio Dias' instrument), but he would also author a history of the São Carlos Theater, first published in 1883 and paying particular attention to the evolution of scenario production.[38]

These technical details about the São Carlos productions point to more than art critics interested in scenery machinery. Benevides was not just an opera enthusiast. He was no less than professor of physics at the Industrial Institute since 1854, becoming its director in 1892. The machinery for the São Carlos theater was installed under his supervision and some of the dynamos were later recycled to illuminate with electric light the classrooms of the Industrial Institute itself.[39] In fact, many of the early electric lighting ventures in Lisbon, such as the Portuguese Company of Electricity (*Companhia Portuguesa de Electricidade*), held Benevides' signature as well as that of many of his Industrial Institute disciples.[40]

While the first director of the Industrial Institute, Vitorino Damásio, had conceived it as a model workshop whose effects were to be extended to the whole of industry, Benevides expected to establish physics as a core field for the new industrial order, surmounting traditional practice/theory divisions. Thermodynamics, optics, and electromagnetism were all taken as crucial subjects to be taught at the Institute to prepare its students for a new world made of steam engines, telegraphic communications, and artificial lighting. Underlining the importance of hands-on learning, Benevides would tour World Exhibitions (Paris 1867 and 1889) and Europe's instrument makers workshops to equip its laboratory as well as the Technological Museum of the Institute.[41]

As demonstration of the central place of instrumentation for the Industrial Institute is its precision instruments workshop, which, as mentioned, was the only one to remain open after the complaints of the Boavista industrialists. While the Institute would struggle until the mid 20th century with the improvised nature of its facilities, a new building was erected exclusively for the precision instruments workshop. This workshop offered a well-designed structure with carefully distributed spaces for the different sections (carpentry, glass works, drawing room, mechanical lathe, testing room) and natural light entering abundantly through the wide lateral windows as well as from the glass rooftop. The precision instruments workshop supplied the national telegraph services with all sorts of electro-mechanic devices; the geographical services with theodolites and levels; the Water Company with meters; public schools with scientific instru-

---

[37] As stated by the critic, "For the large effects of the modern scenic arts it is not enough some nicely painted vistas, one needs as well light combinations impossible to find in our stages before they are replaced with new ones following the demands of scenography new processes." From, "O scenario da opera mephistopheles."

[38] Francisco Fonseca Benevides, *O Real Theatro de S. Carlos de Lisboa*, 1883.

[39] Mario Alberto Nunes da Costa, *O Ensino Industrial em Portugal*, p. 147.

[40] Matos et al., *A Electricidade em Portugal*.

[41] Benevides formed a "collection of machines and mechanisms as well as of kinematic combinations and all the necessary material for the teaching of mechanics and physics," complemented with models of steam engines and other industrial machinery. Francisco da Fonseca Benevides, *Catálogo das Colecções do Museu Tecnológico do Instituto Industrial e Comercial de Lisboa*, 1873.

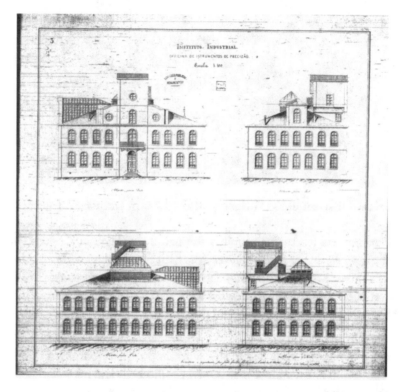

Figure 4.6: **Precision Instruments Workshop of the Lisbon Industrial Institute, c. 1870**
The precision instruments workshop of the Industrial Institute was the only one given a well-designed structure with carefully distributed spaces for its different sections (carpentry, glass works, drawing room, mechanical lathe, testing room) and natural light entering abundantly through the wide lateral windows as well as from the glass rooftop. Instrumentation was key for Fonseca Benevides' vision of an institute training its students in thermodynamics, optics and electromagnetism, preparing them for a technological world of telegraphs, dynamos, and electrical lighting.
© Arquivo Histórico da Economia.

Figure 4.7: **Precision Instruments Workshop of the Lisbon Industrial Institute**, by **Augusto Bobone, 1903**.
The precision instruments workshop supplied the national telegraph services with all sorts of electro-mechanic devices; the geographical services with theodolites and levels; the Water Company with meters; public schools with scientific instruments. All companies in Lisbon involved in the production of electromechanical devices were started by men trained in this workshop.
© Arquivo Fotográfico da Câmara Municipal de Lisboa.


ments. And maybe as important, it was in this workshop that were trained the men that made precision their profession like the afore mentioned Emilio Dias and his manometers for the Gas Factory; António José dos Santos who besides designing steam engines and compasses for solving spherical trigonometry problems, developed breaking systems for wagons manufactured in the Phenix foundry; Maximiliano Hermann who founded a precision instruments company in Lisbon dedicated to electromechanical devices.[42]

In 1888, a new Portuguese Industrial Exhibition conformed the accomplishments of this generation of students of the Industrial Institute. While in 1849 the Portuguese industry had only occupied one large room of the Navy Arsenal in the Boavista area, the 1888 version sprawled through a new major boulevard—the Avenue—that materialized in the cityscape the engineers' dreams of Europeanization of the country. The visitor rambling from one pavilion to another passed through all different categories of industries, each exhibition impressing in his mind the vision of Portugal participating in the same technological world of its European counterparts: typography and other graphic arts in the D. Fernando pavilion; industrial schools at the gallery of António Augusto de Aguiar (recently deceased Minister of Public Works, professor of chem-

[42] "O ensino profissional no albergue nocturno," *O Occidente*, September 11, 1888 and September 21, 1888.

EXPOSIÇÃO INDUSTRIAL PORTUGUESA NA AVENIDA DA LIBERDADE EM 1888

VISTA DO PAVILHÃO DA ENTRADA DA EXPOSIÇÃO

Figure 4.8: **Main entrance of the 1888 Portuguese Industrial Exhibition**.
The exhibition sprawled through a new major boulevard—the Avenue—that materialized in the cityscape the engineers' dreams of Europeanization of the country. Industry and Agriculture Pavilions showcased Portugal as forming part of the European technological world.
© Photo by Mário Novais Studio. Arquivo Fotográfico da Câmara Municipal de Lisboa.

istry at the Industrial Institute and its director during the year 1870); chemical products and precision instruments in the Faria de Guimarães gallery; agriculture produce (wine, olive oil, flour) at the Queen Amélia and Morais Soares pavilions; metal and glass works at the Guilherme Stephens gallery; textiles, furniture, photography, music instruments,…by then Portugal was still predominantly a rural country, but no visitor was to doubt that it was following the path of industrial progress promised by the engineers of the Regeneration and that it was able to produce every possible artifact of the new industrial society.[43]

Confirming the role of the Industrial Institute in this envisioned new industrial order, the illustrated journal *O Occidente* (The West) published parallel to the descriptions of the exhibition a list of its most distinguished students enumerating also their major accomplishments: Miguel Ventura Silva Pinto, demonstrator at the physics and chemistry laboratories of the In-

---

[43] The workers of the different industries paraded in front of the royal stage in which all major societies and associations stood next to the royal family. Here is the description by the *O Occidente*: "these are the labors who form the links of that great chain of work and whose products—united in fraternal embrace—display astounding brightness, amazing variety, surprising strength, and fill our spirit of enthusiasm. They make us cry bravo from the depths of our soul ailing the great orchestra of work that blurs creature and creator, for it also has the power of creation." From, "Exposição industrial portuguesa com uma secção agrícola," *O Occidente*, June 21, 1888.

dustrial Institute; Antonio José dos Santos, head of the pyrotechnics workshop of the Army's maintenance; Caetano Jose de Figueiredo, state supervisor of the engines of the North and East Railway and main engineer of the ice factory; Maximiano Augusto Herrmann, director of an electromechanical instruments company and the first one to install incandescent light bulbs in Lisbon; Emilio Dias, engineer at the Gas Factory; Ignacio Casimiro Alves d'Azevedo, director of the powder factory; Manuel Simões Nascimento, machine builder; Carlos Augusto Pinto Ferreira, director of the machinery workshop of the Navy Arsenal and one of the founders of the socialist cooperative Centro Promotor dos Melhoramentos das Classes Laboriosas (Center for the Improvement of Working Classes).[44]

They were presented as Portuguese industrial heroes who had all passed through the model workshops of the Industrial Institute inaugurated in 1852 by Vitorino Damásio. They had all profited from the new use of gaslight illuminating the institution facilities that converted night time into a space for social emancipation through technical education. In 1888 they could be listed as exemplary men of the Regeneration.

[44] "O ensino profissional no albergue nocturno," *O Occidente*, June 21, 1888.

# CHAPTER 5

# The Colonial Face of Portuguese Engineering

In 1934, the map *Portugal não é um país pequeno* (*Portugal is not a small country*) was presented at the Portuguese Colonial Exhibition, held in Oporto. It showcased Portugal as a strong colonial power offering an alternative vision of the size of the country at the European scale.[1]

Distinct versions of the map were slightly but significantly different: while in the Portuguese version, Portugal is painted in yellow, as all other European countries (image 1), thus bringing to the forefront its position as colonizer, the international versions present both Portugal and the colonies in red (image 2), highlighting the message of a unified strong and large country in the international scene. These two different messages conveyed at the 1934 exhibition embodied Portuguese imperial agenda since the mid 19th-century: the overseas territories were used simultaneously as a strategic asset when Portugal negotiated its place in the international scene and as territories to be exploited and used in a typical colonizer/colonized relationship. The success of both imperial strategies, as we will see, heavily depended on the intense use of Portuguese controlled African territories, both concerning natural and human resources. Technology, and particularly the infrastructure that secured the flows of raw materials, commodities, and people, was at the core of the colonial agenda.

The "material improvements" that summarized engineers' interventions in the national territory also condensed much of Portuguese colonial policy of the 19th century. After the traumatic loss of Brazil in 1822, engineers became main actors in imagining and materializing a new colonial empire now anchored in Africa, namely in Angola and Mozambique, the two largest Portuguese possessions. Before its independence, products from Brazil counted for no less than 2/3 of Portuguese exports and constituted by far the first source of revenue not only of Lisbon merchants but crucially of the Portuguese state. How could Africa, whose main role in the

---

[1] Heriberto Cairo, "'Portugal is not a small country': Maps and propaganda in the Salazar regime," *Geopolitics*, 11, no. 3, (2006), pp. 367–395. The map was imagined by the army officer Henrique Galvão as commissioner of the Oporto Colonial Exhibition. Henrique Galvão participated in the military coup of 1926 that inaugurated dictatorial rule in Portugal (see the following chapters) and was a fierce supporter of the dictator Salazar until the 1950s, when he became critical of the regime. Accused of conspiring against Salazar, he was imprisoned and expelled from the army but he managed to escape in 1959, taking refuge in the embassy of Argentina and having obtained political exile in Venezuela. It was during the exile that Galvão performed a spectacular action against the Portuguese dictatorship—Operation Dulcineia—hijacking the Portuguese passenger cruiser Santa Maria.

Figure 5.1: *Portugal is not a small country*.

These two maps—one for domestic consumption (which stresses the dimensions of the empire) and the other, in English and French, for international use (showing Portugal and the colonies together stressing the unified characteristics of the nation)—are part of the authoritarian regime—Estado Novo (New State)—propaganda presenting the country as a unified imperial country. Created by Henrique Galvão for the 1934 Colonial Exhibition, it was printed by the Secretariado de Propaganda Nacional (National Propaganda Secretariat) and distributed to schools all over the country.
© Public domain.

imperial economy had been of slave supplier of Brazilian plantations and mines, sustain a new empire? How could Africa become a "new Brazil"?[2]

There were many plans drafted by Lisbon political elites on how to transform Angola and Mozambique and build a new imperial block in the aftermath of Brazilian independence. The basic principle was to divert these African territories from the slave business to the production of tropical commodities like sugar, rubber, or coffee, while offering Portugal a protected market for the expansion of its commercial agriculture and industry. The slave trade was indeed formally abolished in the Portuguese empire in 1836 but the strong ties weaved across the South Atlantic between Angola and Brazil in the previous centuries were hard to sever by decree. Brazil would officially stop the import of slaves only in 1851 and finally abolish slave labor in 1888, making the transition a slow one, with south-south relations between economic elites in Brazil and Angola always stronger than south-north ones between Portugal and Angola. Portuguese presence in Angola, as well as in Mozambique, was limited during most of the 19th century, with no strong economic lobby in Lisbon pushing for investments in African colonies. Only in

[2] Valentim Alexandre, *Velho Brasil, Novas Áfricas: Portugal e o Império, 1808–1975*, (Porto: Edições Afrontamento, 2000); A. H. de Oliveira Marques, Ed., *O Império Africano, 1890–1930*, (Lisboa, 2001); campanhas, II, p. 116; Valentim Alexandre and Jill Dias, Eds., *O Império Africano, 1825–1890*, (Lisboa, 1998).

the 1870s did the high rhetoric of imperial restoration translate into concrete initiatives. Not surprisingly, engineers were at the forefront of such undertakings.

## 5.1    ENGINEERING A EUROPEAN COUNTRY THROUGH PUBLIC WORKS IN AFRICA

The main drive for Portuguese colonial enterprises in the 19th century was not the search for new markets by ambitious capitalists of Marxist economic histories, but the voluntarism of the Portuguese state in making the colonies a major factor in the international positioning of the country, namely in relation to its European counterparts. Geostrategic reflections on the threats for a small European nation of the international turmoil caused by the fall of the monarchy in Spain (1868) and the confrontation between the French Empire and Prussia were indeed behind the colonial thought of João de Andrade Corvo, the most consequential of the Portuguese overseas ministers of the Regeneration period.[3] In his widely quoted collection of essays "Dangers" (1870), this engineer, trained at the Army School, professor at the Lisbon Polytechnic School and later its director, pointed at Belgium, the Netherlands, and Denmark as examples of successful small European nations that justified their existence not by military might but by their commitment to "progress."[4] Andrade Corvo, opposing rising trends among Lisbon intellectuals, denied the power of race for defining a nation, a criterion he equated with the classification of animals in a natural history museum, preferring instead to emphasize "civilization" as marker of deserving nations. In the grim international picture of Europe in the early 1870s painted by this minister/engineer, with nations confronting each other and in which Portugal allegedly ran the danger of annexation by Spain, only civilized nations governed by principles of justice, liberty, and, crucially, good administration, could claim their right to existence.

Before serving as minister of overseas and navy, Andrade Corvo had been particularly active in the Portuguese parliament as promoter of a new type of railway lines: economic railways, a phrase pointing at both the economy of means which they were built and their contribution to the national economy.[5] Not all railway lines contributed the same way to place Portugal among civilized nations. Praise should be reserved to those lines that proved how the country was being properly administered. The already-mentioned Southeast line, which connected the capital with expanding cereal production areas of Alentejo, and the Douro line, which connected port wine production areas with the city of Oporto on the coast, were the two exemplary cases of economic lines offered by Andrade Corvo to demonstrate how Portugal was well administered and deserved to exist among European civilized nations.

After the mythical geographic expeditions of the 1870s of Livingstone, Stanley, and Cameron turned the attention of the European nations to Africa, Andrade Corvo would also in-

---

[3] Pedro Calafate, "O pensamento político e geoestratégico de João de Andrade Corvo." *Revista Estudos Filosóficos*, 3, (2017); David Justino, *Fontismo: Liberalismo Numa Sociedade Illiberal*, (Lisboa: D. Quixote, 2016), pp. 250–255.

[4] João de Andrade Corvo, *Perigos*, (Lisboa, 1870), p. 159.

[5] Marta Macedo, *Projectar e Construir a Nação*, p. 200.

clude the colonies in his geostrategic reasoning.[6] Economic lines in Portugal were not sufficient anymore. Colonial railway lines would also become obligatory markers of a civilized European Portugal. In 1876, the infamous King Leopold II of Belgium organized the Brussels Geographic Conference, eventually leading to the partition of Africa among European powers at the Berlin conference later in 1885.[7] Sitting around the negotiations table, the representatives of European nations, showing their voracious appetite for new sources of raw materials and markets, divided the map of the continent according to economic and military might, and put aside historical rights of possession as those claimed by the Portuguese invoking their presence in Africa since the 15th century. Only those countries able to demonstrate their ability to effectively occupy and explore overseas territories could claim the status of colonizer and have a seat at the table.[8]

It was to demonstrate such ability that Portuguese elites founded in 1876 the Lisbon Geographical Society.[9] This society would fund expeditions by Serpa Pinto, Roberto Ivens, and Hermenegildo Capelo, Portuguese counterparts of those British explorers, to assert control of the country African possessions, sparking visions of a Portuguese empire crossing the African continent from the Atlantic coast of Angola to the Indian Ocean coast of Mozambique, a geographer's dream popularized by the Portuguese press as the Rose-Colored Map.[10] Acknowledging the importance of such geographic expeditions, Andrade Corvo was nevertheless convinced these were not enough to count Portugal among European powers claiming rights to occupy Africa. In his condition as minister of navy and overseas, he thus funded two "Public Works Missions" to Angola and Mozambique in 1877.[11] Confirming the voluntarism of the Portuguese state in the colonial enterprise, and employing the common mechanisms for funding the building of infrastructures in the country, the minister used the public deficit to back the international loans funding these African missions.[12]

Applying the same rationale he had used to justify economic railways in Portugal, Andrade Corvo now equated progress in Africa with the construction of roads, telegraph and railway lines connecting the main port cities of the two colonies, Luanda and Lourenço Marques (today Maputo), respectively, to their hinterlands. Portuguese presence was to be felt beyond the historical occupation of the coast, reaching the inner parts of both territories and putting in

---

[6] There is an extensive bibliography on this topic. The classic by Thomas Pakenham, *The Scramble for Africa*, (London: Abacus, 1992) is an excellent base.

[7] Adam Hochschild, *King Leopold's Ghost*, (New York: Mariner, 1999).

[8] Maria Paula Diogo and Dirk van Laak, *Europeans Globalizing: Mapping, Exploiting, Exchanging*, (Palgrave Macmillan), 2016.

[9] Angela Guimarães, *Uma Corrente do Colonialismo Português: A Sociedade de Geografia de Lisboa, 1875–1895*, (Lisboa: Livros Horizonte, 1982).

[10] Cristiana Bastos, "Das viagens científicas aos manuais de colonos: A sociedade de geografia e o conhecimento de África." In *O Colonialismo Português: Novos Rumos da Historiografia dos PALOP (Vila Nova Famalicão: Húmus)* (2013), pp. 321–346.

[11] For details on these missions, Bruno José Navarro Marçal, "Um império projectado pelo "silvo da locomotiva." O papel da engenharia portuguesa na apropriação do espaço colonial africano. Angola e Moçambique (1869–1930)." Ph.D. dissertation Nova University, 2016.

[12] On details on the funding schemes of Portuguese colonial railways see, Hugo Silveira Pereira, "Especulação, tecnodiplomacia e os caminhos-de-ferro coloniais entre 1857 e 1881." *História: Revista da Faculdade de Letras da Universidade do Porto*, 7.1, (2017).

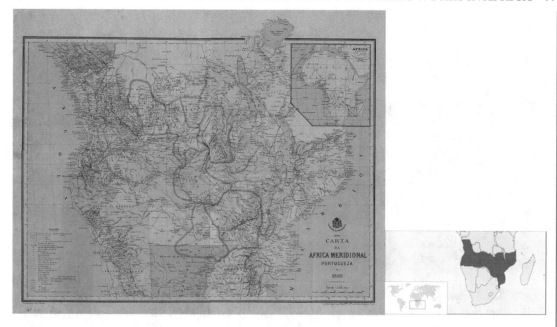

Figure 5.2: **The *Mapa cor-de rosa* (rose-colored map).**
The rose-colored Map—claiming a continuous strip of Portuguese sovereignty between the western (Angola) and eastern (Mozambique) coast of Africa—was an attempt by the Portuguese government to assert its *status* as major colonial power after the Berlin Conference (1884-85) and create the necessary conditions to build a west-east (coast to coast) railway line. However, this Map and Cecil Rhodes' north-south axis project of connecting British possessions in Africa from Cairo to Cape Town could not co-exist. The clash with British imperial plans relegated Portugal to second tier European power and unleashed a structural long-term national crisis.
© Public domain. Collection of the Sociedade de Geografia de Lisboa.

contact populations of the interior with international markets. The transportation and communications infrastructure should transform cultivation of crops for auto consumption and local markets into large scale production of tropical commodities, namely sugar, rubber, and coffee. As in Portugal, engineers' infrastructure was trusted with the task of redeeming Angola and Mozambique from its perceived state of backwardness:

> We can't keep leaving isolated, as when our African colonies were parks for the production and breeding of slaves and not much else. Today the world belongs to work and not to indolence; the land is for men and no one may take from civilization what it belongs to her. It is necessary to produce, to produce in large scale, where nature concentrated its production forces… Not doing anything in favor of civilization and humanity and demand respect from others; occupy vast regions of the world without

*Expedição dos primeiros estudos do Caminho de Ferro entre Luanda e Ambaca*
Nesta gravura vêem-se : *ANGELO DE SARREA PRADO, iniciador, que efectuou os trabalhos técnicos e respectivo ante-projecto e Capitão J. M. BARREIROS ARROBAS, conductor de obras públicas*

Figure 5.3: **Preparatory works for the Luanda-Ambaca (Angola) railway (c. 1887).**
Following the Berlin Conference (1884-85), which established the principle of imperial sovereignty through material occupation of the land, Portuguese colonial policy was reorganized in Angola and Mozambique toward the building of infrastructures, namely railways. The Portuguese caption refers only the names of the European engineer and the auxiliary technician, ignoring the African workers that participated in the public works expedition.
© Public domain. In *Gazeta dos Caminhos de Ferro*, no. 1110, 16 February 1934.

accepting the responsibility weighing on us; these are the errors that trouble reason and enfeeble the physical and moral faculties of our country."[13]

The ideology of work and production didn't just inform engineers' projects for transportation infrastructures crossing Portuguese territory or for increasing the skills of Lisbon industrial workers, as explored in the two previous chapters. If work and production were the new definers of civilization, the burden of the Portuguese in Africa—according to Andrade Corvo—was to divert local populations from indolence to productive work. Roads, telegraphs, and trains would contribute to such end not only through the large workforce mobilized for their construction, but mainly by connecting the interior of Angola and Mozambique with international markets thus promoting the cultivation of commodity crops. The materiality of engineers' infrastructure, not history or tradition, confirmed the European civilized identity of the Portuguese and thus their right to occupy territories in Africa. Public works materialized the Portuguese "civilizing

---

[13] João de Andrade Corvo, *Estudos sobre as Províncias Ultramarinas*, vol. I, (Lisboa, 1881–1887), p. 212.

mission" in Africa, as defined by the engineer.[14] It was thus appropriated that Andrade Corvo named the enterprise as Public Works Missions.

## 5.2  NATIONAL PRESTIGE AND PROFESSIONAL STRATEGIES

The technological landscape dreamed by Andrade Corvo in Africa was decisive to keep Portugal as an active actor in the European and international scene. Nevertheless, Portuguese engineers' commitment to the colonial enterprise should be perceived beyond notions of patriotic duty, to include as well the dimension of a professional imperative. It was not lost among the members of the profession, that the effective occupation of the African territories through public works created a new market for technical expertise, advancing young engineers' career prospects. Africa offered young graduates in civil engineering from the Army School the chance to work on major infrastructures and to acquire a specific and valuable training, which later on could be reinvested in their careers back home. In 1899, the *Journal of Public Works and Mining*, the mouthpiece of the Portuguese Association of Civil Engineers, published an important document known as the *Alvitres* (*Suggestions*), in which Portuguese engineers were invited to present their main matters of concern and to present suggestions to be undertaken by the government.[15] Concerning the issue of employment (suggestions 9 and 10), the document emphasized both the importance of the African territories as a promising job market and, once again, the role played by engineers as the artisans of progress and sentinels of national pride in the overseas territories: "When colonial nations do effectively want to take possession of their territories they send their engineers overseas."[16]

The importance of the colonial topic for the engineering profession can be indeed asserted by looking at the articles on public works in the African colonies published by the *Journal of Public Works and Mining*. As it is clear, apart from the period of World War I, during which the journal had some problems concerning its publication, the number of articles on public works in Africa draws a steady increasing curve. These articles ranged from the short news genre to the public lecture and to the extensive *memoir*, which included a general approach to the topic, always stressing the Portuguese "civilizing mission," followed by a technical report on the specific issues related to a specific infrastructure. Engineers were changing the face of the Portuguese African territories, by allegedly "domesticating" African wilderness and duplicating the technologically driven agenda of the mainland in the colonies. They were artisans of national pride in the world arena, by offering Portugal the tools to secure its empire.

---

[14] Miguel Bandeira Jerónimo, *The "Civilising Mission" of Portuguese Colonialism, 1870–1930*, (New York: Palgrave MacMillan, 2015).

[15] See M. P. Diogo, "Industria e engenheiros no Portugal de fins do século XIX: o caso de uma relação difícil," *Scripta Nova*, 2000, 69(6) (http://www.ub.es.geocrit/sn-69-6.htm).

[16] "Exposição" (Acta da sessão de 6 de Maio de 1899), *Revista de Obras Públicas e Minas—R.O.P.M., Journal of Public Works and Mining* (1899), pp. 353–354; pp. 382–383.

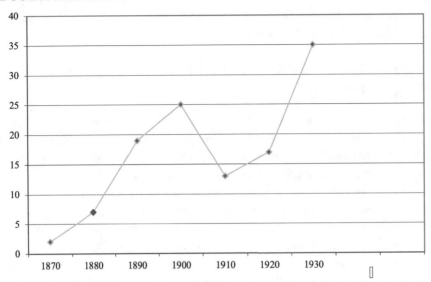

Figure 5.4: Articles about African public works published in the *Journal of Public Works*. Source: *Revista de Obras Públicas e Minas* (*Journal of Public Works and Mines*) (1870–1936).

After linking the coast and the inner territories by telegraph and telephone, with their lines running side by side with the railway lines (…) the Portuguese spirit will flow through the radio waves from the top of the Peninha [a small peak in a village near Lisbon] until Mozambique. It will be as if the good old [imperial] days were back with Portugal controlling Africa from coast to coast.[17]

Although telecommunication networks played a major role in positioning the country in the international agenda, railways were the dominant colonial theme of the journal.[18] The seminal *memoir* on the railway line from Lourenço Marques to the Transvaal, written by the engineer Joaquim Machado and published in 1882, became indeed a model for other reports. This was no less than the report of the Public Works Mission to Mozambique launched by Andrade Corvo in 1877.[19] The first part of the text had the suggestive heading "The need to build up railways to civilize Africa." The historical and political context was used to justify the construction of a railway, which was presented as embodiment of civilization rather than a mere technological

[17] The quest referred to above for the rose-colored Map—a Portuguese imperial dominion from the Atlantic to the Indian ocean coast—was again afoot this time by virtue of the proficiency of the national engineers. "Relatório e contas da associação dos engenheiros civis Portuguezes com referencia ao ano de 1904," *R.O.P.M.*, (1905), pp. 421–423; pp. 62–64.

[18] On telegraphs see, Ana Paula Silva and Maria Paula Diogo, "From host to hostage: Portugal, Britain and the Atlantic telegraph networks." *Networking Europe: Transnational Infrastructures and the Shaping of Europe, 1850–2000*, (2006), pp. 151–86.

[19] The other mission to Angola, led by the engineer Rafael Gorjão, studied the viability of a railway line connecting Luanda to Ambaca, in the interior of the territory.

Articles Published in the *Journal of Public Works* on African Topics

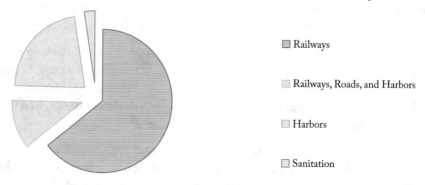

Railways

Railways, Roads, and Harbors

Harbors

Sanitation

Figure 5.5: Articles published in the *Journal of Public Works on African topics*. Source: *Revista de Obras Públicas e Minas* (*Journal of Public Works and Mines*) (1870–1936).

device. Reworking the arguments previously articulated by Andrade Corvo, Machado described how

> a few years ago, knowledgeable men of the African wilderness became aware of the fundamental help provided by railways in transforming and civilizing this immense continent, for the most part still in a barbaric stage, and almost exclusively populated by a primitive and savage race devoted to hunting and war and living in complete darkness.[20]

Railways were considered the main tool for the integration of African territories and populations as resources and producers in international markets. But Machado offered railways a moral dimension as well, serving simultaneously to assert the degree of civilization of the Portuguese, confirming that they were indeed Europeans—an obsession of Portuguese engineers, always self-conscious of their status—while demonstrating the primitivism of the native population.[21] In other words, railways in Africa made the Portuguese white Europeans, and the local populations black Africans who lived in "complete darkness."

---

[20] Machado, "Memórias," 2.

[21] This is also the famous topic of Michael Adas, *Machines as the Measure of Men. Science, Technology and Ideology of Eastern Dominance*, Ithaca, 1989. See also Maria Paula Diogo and Dirk van Laak, *Europeans Globalizing: Mapping, Exploiting, Exchanging*, Palgrave Macmillan, 2016.

In his report, Machado compared old strategies (military actions and religious indoctrination by missionaries) to modern means (railways), offering the "Gee-whiz" effect[22] as the best way to dazzle native populations and enhance their desire for progress and civilization.

> In the century of steam engines and electricity Europe does not have to use old methods to civilize Africa. A gun, a sophisticated fire engine, a steam engine, a large road, a railway, the whistle and movement of an engine, etc., produce in the inhabitants of Africa a deeper stimulus to their intellectual development than masses and sermons preached by the most eloquent missionary.[23]

These naïve general considerations on how the simple presence of modern technology converted local populations to the supposed benefits of colonialism were followed by detailed tedious descriptions of the enterprise with abundance of numbers, exemplifying to fellow engineers how to describe an economic colonial railway. Machado produced a financial calculation of needed investment and future profits and accessed the technical aspects of the railway: the configuration of the railway, its stations and the distance between rails: "The railway is 1.08 m large between rails, approximately 70,000 m long and is divided into four sections": from Lourenço Marques to the right bank of river Matolla; from the right bank of the Matolla to Saguans lowlands; from here to the Umchamachabas plains; and then to the border. The railway included nine "works of art": three viaducts all in stone, and six bridges, three in stone and three in stone and iron. Other features, such as railway crossings, main and secondary railway stations, fences, workshops, and houses for railway keepers, were also mentioned. Considering that two trains would circulate in opposite directions each day, the engineer produced an estimate of needed equipment: 3 locomotives, 3 passenger carriages with 58 seats each, and 6 freight wagons. Finally, Machado calculated the workforce costs adding to the natives used in non-specialized tasks the importation of qualified European workers.

The actual construction of railway lines in Mozambique, as well as in Angola, got delayed by the usual funding difficulties, the Portuguese state finally ending by guaranteeing the return on investment that would make construction viable. But in spite of the delays, the truth is in 1906, 783 km of railways were already built up and working in Angola and Mozambique, and around 400 km more were in the process of construction. In 1914, on the brink of World War I, Portuguese African railway lines were no less than 1,600 km long.[24] Reports and projects on public works, particularly the technical and management guidelines and procedures for building railways as the one produced by Machado, should thus be taken as critical historical elements to understand the scale of changes imposed on African colonies. These reveal how engineers put forward a new geography and hierarchy of economic spaces determined by colonial visions of

---

[22] The term "Gee-Whiz" has been used by historians of technology to characterize the joint attitude of amazement and veneration in response to a technical object. It was used by Melvin Kransberg in the "Introduction" to the book *The History and Philosophy of Technology*, (Chicago, 1979), on page 14 to classify the attitude of those which viewed technology and its history as an epic succession of wonders. The term "O-gee" effect can also be used with the same connotation.

[23] Machado, "Memórias," 5.

[24] A. Castilho, "As grandes linhas do caminho de ferro," *R.O.P.M.*, (1914), 45, pp. 638–640.

Figure 5.6: **Building the Lourenço Marques-Transvaal railway (c.1887).**
This line connecting the port of Lourenço Marques (Maputo) with the Transvaal mines was the first ordered by the Portuguese government. Notice the use of the locomotive to assert European manhood and superiority.
© Public domain. In C. S. Fowler, Álbum *Views of Lourenço Marques (Delagoa Bay) and Transvaal Railway.*

Africa as producer of commodities for the metropole, while ignoring local ways of appropriating and using the land.

Railways and roads had indeed major implications for colonial historical dynamics, but it should be clear that these were never as smooth as idealized by engineers' memoirs. The city of Lourenço Marques (Maputo), for example, became a major port for the export of riches from the African interior due to the new railroad first projected by Machado. But the diamonds and gold came from Transvaal mines under British control, with no obvious advantages for the hinterland regions of Mozambique under Portuguese control. In fact, the railway proved to be an important factor in depleting Mozambique of his agriculture workforce with young male migrants leaving seasonally the Portuguese colony for the Transvaal mines.[25] The Portuguese colonial state did profit from such arrangements, retaining a percentage of the salary of the exported workforce as well as the export dues of the Lourenço Marques port. But this was far from the civilizing effects promised by Andrade Corvo's economic railways.

---

[25] Joel Maurício das Neves, Economy, society, and labor migration in central Mozambique, 1930–c.1965: A case study of Manica province, Ph.D. dissertation SOAS, London, 1998; José Negrão, *Cem Anos de Economia da Família Rural Africana*, (Maputo, Promedia, 2001); José Capela, *O Álcool na Colonização do Sul do Save*, (Maputo, 1995).

Figure 5.7: **African workers in the Lourenço Marques-Transvaal railway line**.
The Portuguese colonial projects under the whitewashing European label of "civilizing mission"
imposed violent physical, social, and economic changes across African territories. Railways and
other technological infrastructures designed by engineers transformed both the African landscape *per se* and its use, by carving the way to plantations, mining, and trade outposts. Forced
labor regimes mobilized local African populations who built the new colonial infrastructure on
their backbone.
© Public domain. In C. S. Fowler, Álbum *Views of Lourenço Marques (Delagoa Bay) and Transvaal Railway).*

And more important than pointing how concrete historical dynamics diverged from the
engineers' civilizing mission promises is to acknowledge the structural problems of such rhetoric.
The emphasis on the virtues of work that had led in the metropole to the creation of the Lisbon
Industrial Institute meant in the colonies the proliferation of forced labor regimes. Putting an
end to the old slavery networks had been a major topic that justified Portuguese initiatives in the
continent, but colonial officials always had trouble to find out volunteer work to cultivate cash
crops or, not surprisingly, to build the infrastructures designed by engineers. The decree of 1899
that would regulate labor practices in the Portuguese empire for the following decades explicitly forced natives to be actively employed giving colonial authorities the power to use physical

coercion to make them work in plantations or public works in case of refusal.[26] As brutally summarized by the minister and governor of Mozambique António Enes, colonial authorities had to zeal over the natives' "work habits." Apparently, the stimulus of African intellect produced by the simple vision of the locomotive was never a sufficient argument to convert Africans to the joys of civilization.

Not surprisingly, many African rulers of interior areas of Angola and Mozambique, those that the new transportation and communication infrastructure aimed to reach, didn't welcome the alleged superior civilization Portuguese brought with them. Bloody long wars were waged between the Portuguese and local powers until the territories under effective imperial control matched those of the maps drawn at the Berlin conference of 1885.[27] These so-called "pacification campaigns" of Angola and Mozambique lasted for more than three decades until the aftermath of the World War I, and their grim history eloquently demonstrates the inadequacy of conceptualizing empires as overwhelming external powers imposing their will on backward populations. The narratives of such campaigns suggest instead the fragility of Portuguese colonial presence, constantly menaced by historical actors that contested the divisions civilized/barbarian of official imperial rhetoric.[28]

At first sight, everything seems in perfect order in the engineer's report: railways and roads communicate productive areas of the hinterland with cities on the coast, and from here the riches of Africa—coffee, rubber, or sugar—are shipped to the major ports in Europe. The overwhelming colonizing power of technology offers no discussion. But a more careful reading reveals the fears and fragility of the whole enterprise. It was out of a sense of national vulnerability that Andrade Corvo, as he explained in his essays significantly titled "Dangers," launched the Public Works missions that materialized Portuguese presence in Africa. Such missions were the ones that would prove that the country was in fact a "civilized" European nation thus guaranteeing, in his vision, Portugal's own independence. Once again, engineers equated their own work, their infrastructure projects, with the destiny of the nation, portrayed as in constant danger of extinction.

The engineering profession now benefited from an extend new field of operations and its elements could also claim they were the men who made Portugal participate of the European "civilizing mission" of Africa. But only to confirm the contested civilized status of Portugal, and the relevance of vulnerability to grasp the nature of Portuguese colonization in Africa, in

[26] Miguel Bandeira Jerónimo and José Pedro Monteiro, "'Das dificuldades de levar os indígenas a trabalhar': O sistema de trabalho native no império colonial Português," in *O Império Colonial em Questão: Poderes, Saberes e Instituições*, Miguel Jerónimo, Ed., (Lisbon: Edições 70, 2012), pp. 159–196.

[27] René Pelissier, *História das Campanhas de Angola: Resistência e Revoltas, 1845–1941*, (Lisbon: Estampa, 1986) and *História de Moçambique: Formação e Oposição, 1854–1918*, (Lisbon: Estampa, 1987–1988).

[28] On the importance of imperial vulnerability to understand colonial dynamics see, C. A. Bayly, *Empire and Information: Intelligence Gathering and Social Communication in India, 1780–1870*, (Cambridge University Press, 1999). Also, Ann Laura Stoler, "Rethinking colonial categories: European communities and the boundaries of rule." *Comparative Studies in Society and History*, 31.1, (1989), pp. 134–161. On the notion of vulnerability applied to the Portuguese empire in Africa see, Ricardo Roque "The razor's edge: Portuguese imperial vulnerability in colonial Moxico, Angola." *The International Journal of African Historical Studies*, 36.1, (2003), pp. 105–124.

Figure 5.8: **Behold, Civilization!**
This cartoon authored by the famous Portuguese caricaturist Raphael Bordalo Pinheiro shows a portrait of Alexandre Serpa Pinto, iconic Portuguese African explorer, on the upper left side. Below, African natives in awe looking at a document presented by Serpa Pinto that reads "Civilization: Railways." Pinheiro was a master ironist unveiling the inconsistencies of the Regeneration period. The apparent simple logic of the European explorer subjugating African populations through the invocation of civilization in the form of railways is subverted by suggesting the naivete of Europeans assuming unproblematic subjugation by Africans. The British Ultimatum delivered in January 1890 demanded Portuguese retreat from the areas claimed by Serpa Pinto ending the dreams of the rose-colored map and opening a long term political crisis in Portugal.
© Public domain. In *Pontos nos ii*, no. 237, 9 January 1890. Available at http: //hemerotecadigital.cm-lisboa.pt/OBRAS/PONTOSNOSII/1890/N237/N237_master/ N237.pdf.

Figure 5.9: **Joao de Andrade Corvo,**
Andrade Corvo, trained at the Lisbon Polytechnic School and the Army School, exemplified
the political involvement of engineers in the Regeneration regime becoming a member of parliament, diplomat and minister. He importantly articulated the rationale for Portuguese imperial
ventures. It was out of a sense of national vulnerability that Andrade Corvo, as explained in his
essays 'Dangers', launched the Public Works Missions in Africa. These should demonstrate that
the country was a "civilized" European nation thus guaranteeing, in his vision, Portugal's own
independence among rival powers. Engineers like Corvo repeatedly equated their own work,
their infrastructure projects, with the destiny of the nation, portrayed in constant danger of extinction.
© Public domain. In *Revista contemporanea de Portugal e Brazil*, September 1860.

1890 the British empire imposed an *Ultimatum* on Portugal. It demanded the withdrawal of
Portuguese troops from Mashonaland and Matabeleland (current Zimbabwe) and from Shire-
Nyassa (current Malawi) putting a definite end to Portuguese ambitions of controlling the interior regions of Africa between Angola and Mozambique, the famous rose-colored map that
connected under Portuguese rule the whole area from the Atlantic coast of Angola to the Indian Ocean coast of Mozambique. The Portuguese government accepted British conditions and
ceased hostilities against the Makololo, which not only confirmed the status of Portugal as second tier European power as it also placed the country in an equal step with the "primitive savage
races" it claimed to civilize.

CHAPTER 6

# The Modernist Engineer

After the celebratory mood set by the 1888 Industrial Exhibition and the high rhetoric associ-ated with Portuguese colonial undertakings in Africa, the 1890s were crisis years in Portuguese history. The British *Ultimatum* of 1890 put an abrupt end to the dream of the rose-colored map and thus to imperial ambitions of occupying present-day Zimbabwe and connecting the Por-tuguese African colonies of Angola and Mozambique. This was followed in 1891 by a profound economic depression that would lead to the final demise of the Regeneration regime, the over-throw of the monarchy, and the inauguration of the Republic in 1910. Portuguese economy felt acutely the global recession started with the Barings Bank in London near insolvency, which led to the sovereign debts crisis similar to the more recent 2008 "Great Recession." Portugal was expelled from the gold standard, defaulted on its debt, and had no access to international credit from 1892–1902. This ended the Regeneration system of funding new infrastructures by running budget deficits and issuing state bonds. Major public works, such as the expansion of the Lisbon port that had promised to transform the city into a major Atlantic hub competing with Rotterdam and Antwerp, were abruptly halted.

Considering how central were roads, telegraphs, and railways for engineers' *esprit de corps*, the national crisis meant a profound rethinking of the place of engineers in Portuguese society. But what was first a major menace to the profession would be transformed into an opportunity for the expansion of engineers' presence in the life of the country. The Republican insistence on training elites for guiding the nation, developing national resources, and increasing investment in African colonies, would totally redefine what it meant to be an engineer in Portugal. To their previous role of managing the territory as builders of transportation and urban infrastructures, typically working for the state, they would add in the early 20th century that of managers of industries transforming the territory, mostly employed by the private sector. And, as in previous centuries, more than just adapting to a new adverse context, engineers were active historical actors striving to build a new national context through their technological endeavors.

## 6.1 TRANSPORTATION AND FIXATION, OR THE ENGINEER'S NEW IDENTITY

The trajectory of Ezequiel de Campos (1874–1965) offers a unique vantage point to start explor-ing this historical dynamic. Not being able to find an attractive job after finishing his formation as Mining Engineer at the Oporto Polytechnic Academy, he joined the colonial administration of the small African Atlantic island of São Tomé where he stayed from 1899 until 1911 as head of the local public works.[1] This confirms the new role of colonies as an important job market for Portuguese engineers mentioned above. Relegated to such a faraway post, Campos was soon confronted with the conservation problems faced by the tropical island, then the first-world producer of cocoa. The unrestrained clearing of the rain forest by cocoa growers to increase their cash crop area drastically changed the landscape and, allegedly, the island's climate.

Drawing on British and French colonial literature, but also on North American conservationist proposals, Campos presented a plan for the recovery of São Tomé based on the "scientific reforestation of the island." New clearings were to be banished; rows of trees planted along the shore forming natural wind barriers; cocoa trees cultivated next to shading trees to protect them from direct sunlight. But forest policies were deemed insufficient to recover soil fertility, which was to be enhanced not only by the proper use of manures, but also, quite surprisingly for a tropical island, by irrigation. Thus, Francis Newell, the U.S. Bureau of Reclamation commissioner responsible for the irrigation works in the semiarid lands of the American West, found his way into a book on the equatorial island of São Tomé. As stated by Campos, "the time is fast approaching when no water will flow through the lower parts of the river, except for floods."[2]

The tropical experience familiarized Campos with American conservationist literature, which would also provide him the principles for the reorganization of Portugal. Inspired by the large-scale ambitions of American Progressive era engineers, Campos proposed to apply the principles of rational management of natural resources not only to the colonial island but to the metropole as well.[3] In fact, it didn't take too long for São Tomé to become too small for Campos plans. He had no problems in shifting Africa for Europe for, as he stated, the "biggest problem of Portugal…is the colonization of Portugal itself."[4] Large areas of the country remained out of the national economy, either due to the alleged backwardness of local populations practicing a subsistence economy or, even more dramatic, to the scarce settlement of the land, namely in the south. The fall of the Portuguese Monarchy by the Republican coup of 1910, that culminated the historical crisis of 1890, was the perfect moment to present his project of refunding the nation on a scientific basis: The crisis opened up the realm of possibilities for imagining the country beyond the infrastructure policies of the 19th century. Campos' return from Africa as national deputy was followed by intense political activity to put forward an ambitious program combining American conservationist movement with the Portuguese tradition of internal colonization projects that had long ago denounced the abandonment of large tracts of land.[5]

---

[1] For Campos' experience in São Tomé see Ezequiel de campos, *A Revalorização Agrícola de S. Tomé*, (V. N. de Famalicão: Tipografia Minerva, 1921); *Modificação do Ambiente das Ilhas de S. Tomé e Príncipe*, (Lisboa: Sociedade de geografia de Lisboa, 1956).

[2] E. Campos, *A Revalorização*, p. 194.

[3] Donald Worster, *Rivers of Empire: Water Aridity, and the Growth of the American West*, (New York: Oxford University Press, 1985), p. 155.

[4] Ezequiel de Campos, *Pregação no Deserto*, (Porto: Lello & Irmão, 1948), p. 12.

[5] On the long tradition of Portuguese internal colonization projects, see Manuel Villaverde Cabral, *Materiais para a História da Questão Agrária em Portugal—Séc. XIX e XX*, (Porto: Inova, 1974). For the American conservationist movement during the Progressive era see namely, Samuel Hays, *Conservation and the Gospel of Efficiency: The Progressive Conservation*

Figure 6.1: *A Conservação da Riqueza Nacional by Ezequiel de Campos (1913).*
The engineer Ezequiel de Campos presented in his thick volume *The Conservation of National Wealth* a full program to save Portugal from the crisis started in 1890. To properly manage soil, water, forests and minerals, meant to properly manage Portuguese society. Maps, as the one represented in the cover, demonstrated the alleged irrationality of Portuguese population distribution. The central issue of Campos' program, inspired by American reclamation of the West, was to irrigate the dry regions of the south and colonize them with people from the humid northwest. Campos' proposals constituted the backbone of Portuguese engineers' visions for the country in the twentieth century.
© Public domain.

In 1911, Campos presented to the National Assembly a law on unreclaimed land, which explicitly recognized as guiding principles the conclusions approved in the American National Conservation Congresses of 1909 in Seattle (Washington) and of 1910 in Saint Paul (Minnesota).[6] The future of the nation was presented as direct result of the conservation of natural resources: to properly manage soil, water, forests, and minerals, meant to properly manage Portuguese society. In his widely quoted *A Conservação da Riqueza Nacional* (The Conservation of National Wealth), a thick volume of more than 700 pages published in 1913, Campos detailed his whole program to save Portugal from its apparent inexorable decline. Social problems were depicted as a series of maps of Portugal accounting for the density of population, the numbers of emigration, and rainfall distribution.[7] The maps measured the deficit of national resources and showed the irrationality of Portuguese population distribution. People were migrating from the densely populated regions of the rainy northwest to Brazil and the United States, while the dry lands of the south, the Alentejo region, could be considered as a no-man's land. The central issue of Campos' program was almost self-evident: to colonize the south of the country with people from the northwest.

Such move would only be possible with a drastic change of the landscape. The semiarid southern lands were to be irrigated by major hydraulic works enabling settlers coming from the north to practice a profitable agriculture. American reclamation heroes like Powell, Mead, or Newell would be proud of the Portuguese engineer who claimed that "the problem of the national destiny is intrinsically dependent of the farming of our land, and this is only possible by large scale irrigation of our semiarid region."[8] The large and unproductive estates of the south, the mythical *latifundia*—should be divided in order to sustain a virtuous community of farmers: "The Republic was born for every Portuguese, offering a homestead to anyone willing to cooperate with the strengthening of the Nation. There is no Republic while a single large landowner subsists." Campos considered that "there is almost no difference between the social condition of the Far West and our Alentejo: both are lands to colonize, the only disparity being that the free lands from the Rockies…to the Pacific shores are mainly Public Domain while among us the Alentejo is private property."[9] The analogy went even further: "May tomorrow the Alentejo be a promise land, a new California."[10]

The greening of the Alentejo by hydraulic works would be complemented by other measures taken from American conservationists' shopping list and adopted to Portuguese condi-

*Movement, 1890–1920*, (Cambridge, MA: Harvard University Press, 1959); Worster, *Rivers of Empire*; Donald J. Pisani, *Water, Land and Law in the West: The Limits of Public Policy, 1850–1920*, (Lawrence, KS: University Press of Kansas, 1996).

[6] Campos, *Pregação*, pp. 20–22.

[7] From 16,000 in 1900, the number of Portuguese emigrants rose to 31,700 in 1910. From 1900–1930 the total flux of emigrants was of some 900,000 people. The total Portuguese population was 5,446,760 in 1900 and 6,802,429 in 1930.

[8] Campos, *Pregação*, p. 17.

[9] E. Campos, *A Conservação da Riqueza Nacional: A Grei, os Mineraes, a Terra, as Matas, os Rios*, (Porto, E. de Campos, 1913), p. 460.

[10] Campos, *Pregação*, 18. Ezequiel de Campos conducted his own colonizing test in Alentejo by buying some 42 hectares of land and occupying it with settlers from the northern region. In his memories he narrates all the difficulties that would finally put a sad end to the experiment. *Pregação*, pp. 33–36.

tions. Forestation, namely, would transform barren mountain slopes into profitable properties, provide a defense against soil erosion, protect riverbanks, prevent floods, and, more ambitiously, change the Portuguese climate. Campos' technological garden, combining dams and forests, also included mines (coal, iron, and copper) and an expanded transportation network. More importantly, it would be crossed by an electrical grid carrying the energy produced by Portuguese rivers.[11] The historical interest of a character like Campos for our argument is how his reformist agenda of the early 1910s would constitute much of Portuguese engineers' agenda for the upcoming decades.

Campos' claims for the building of a national electrical grid presented in conferences, newspapers, and parliamentary sessions went hand in hand with his own explorations of the main Portuguese rivers. He traveled along the Douro, Cávado, Mondego, Guadiana, and Tagus, the main Portuguese rivers, in search for the hydroelectric power to support the industrialization of the country. Making once again explicit his African references, he confessed to have endured more tribulations in Portuguese remote villages than in his African experience in São Tomé.[12]

After six years of wandering through allegedly unexplored Portuguese territory, sleeping among local villagers, and carrying his own surveying instruments, Campos' efforts were recognized by the Republican government, which appointed him head of the Hydraulic Studies Brigade created in 1918. The moment was indeed appropriate, for the World War I had dramatically revealed the incapacity of Portugal to rely on its own resources, with energy and food shortages leading to popular riots in Lisbon and across the country. While new irrigation areas would increase the national capacity for food production, hydroelectricity would provide a national source of energy freeing the country from its foreign dependency of coal and thus closing the industrialization gap with more developed nations. As in all other proposals by engineers of previous years, comparisons with other European, and now also North American, countries were obligatory.

For the country to thrive, rivers, together with the national soil resources, were to be scientifically managed. This was the solution Campos offered to the "Portuguese Problem" (his own words) and that would guide much of Portuguese engineers' undertakings until the last quarter of the 20th century. Engineers were to become main protagonists in the urgent task of replacing the so-called old policy of "transportation," based on commerce, by the new policy of "fixation"—the development of the country's own possibilities through industry and agriculture. This influential formulation was due to António Sérgio, the public intellectual responsible for publications such as *Pela Grei* (For the People) or *Seara Nova* (New Harvest) that gathered much of the Republican elite that promised to guide the country out of the path of perceived decadence. In these magazines, pedagogy, fiction, philosophy, and poetry shared space with economy, agronomy, and engineering, with Campos putting forward his hydraulic arguments next to Raul Proença's political essays, Aquilino Ribeiro's novels, and António Sérgio's proposals

---

[11] On Campos's proposals for a National Electrical Grid see E. Campos, *Problemas Fundamentais Portugueses*, (Lisboa Revista Ocidente, 1946), pp. 150–151.

[12] Campos, *Pregação*, pp. 61–63; Campos, *Problemas Fundamentais*, pp. 151–152.

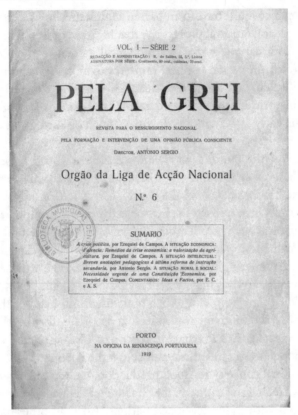

Figure 6.2: *Pela Grei (1919)*.
The Republican Intellectual António Sérgio promoted journals like *Pela Grei* (For the People)
or *Seara Nova* (New Harvest) to forward his vision of 'fixation' replacing 'transportation' as the
solution to the country's perceived crisis. Pedagogy, fiction, philosophy, and poetry shared space
in the journal with economy, agronomy, and engineering, namely Ezequiel de Campos' hydraulic
projects.

for experimental educational practices. Portuguese cultural historians have justifiably paid great
attention to all those figures, and namely to Sérgio the leader of the group. It is surely fascinating
that Sérgio wanted to count with the presence of the engineer Campos from the very beginning
of his project, inviting him to write in the papers of his journal next to other intellectuals. But
maybe more important for our purpose is signaling how other engineers would follow Campos,
positioning themselves right next to humanists in imagining futures for the country.

## 6.2    REPUBLICAN EDUCATION AND THE TECHNICAL INSTITUTE: *TECHNIK* COMES TO PORTUGAL

The reorientation of national life from "transportation" toward "fixation" had a significant institutional translation in engineering education. While the combination of the Polytechnic School and the Army School dominated Portuguese engineering during the Regeneration—the quintessential "transportation" years—the Lisbon Industrial Institute—renamed in 1911 as Superior Technical Institute (IST) (from here onward Technical Institute)—would assume the role of main engineering institution for the 20th century and the new era of "fixation." The military ethos that dominated a corps owing its reputation to the transportation infrastructure built during the 19th century, would give way to a new industrial/managerial identity, with cohorts of new engineers employed in industries transforming the national soil into a resource producing fertilizers, cement, paper pulp, metals, and, of course, electricity.

It is certainly not a coincidence that the main protagonist of such institutional transformation of redirecting engineers to the Portuguese soil was a mineralogist, Alfredo Bensaúde, professor at the Industrial Institute. Importantly, as we shall see, he had completed his engineering training in Germany at the Technical School of Hannover, the Mining School of Clausthal, and the University of Gottingen. Already in 1892, the year Portugal lost access to international credit, Bensaúde proposed to restructure technical training,[13] identifying the problems afflicting his institution: the lack of professional "hands-on" experience of most of the faculty; the few theoretical papers, articles or books written by the faculty; the theoretical minded classes, precisely the opposite of a true technical teaching based on everyday practice; and the little attention paid to technical drawing, "in a way the written alphabet of the technician."[14] More theory and more practice, or better said, a new relation between the two, reinforcing each other.

Bensaúde proposed a set of five technical courses dealing with building, mechanics, electricity, chemistry, mining, and commerce. The proposal was harshly criticized by its peers, mainly because it was considered too "European," i.e., not supported by Portuguese tradition. Bensaúde, who was indeed an "europeanized" engineer, argued that "good examples should be followed no matter where they came from and that a good school was always, in a way, the result of international influences."[15] One should of course keep in mind that the Portuguese tradition referred by the detractors had been molded on the example of French technical education system. The debate was thus a question of determining if to become a good Portuguese engineer one should look at the French *Polytechnique* or German *Technische Hochschulen*.

Although Bensaúde's plan was not immediately implemented, some changes were soon introduced in the *curricula* of the Industrial Institute in order to close in the alleged gap between industrial needs concerning specific skills and industrial teaching. To finish their train-

---

[13] Alfredo Bensaúde, *Projecto de Reforma do Ensino Technologico para o Instituto Industrial e Commercial de Lisboa*, (Lisboa: Academia Real das Ciências, 1892).

[14] Bensaúde, *Projecto*, pp. 10–23.

[15] Alfredo Bensaúde, *Notas Histórico-Pedagógicas Sobre o Instituto Superior Técnico*, (Lisboa: Imprensa Nacional, 1922), p. 5.

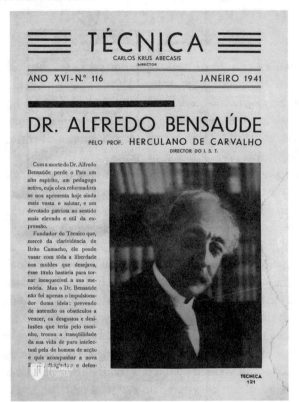

Figure 6.3: **Alfredo Bensaúde (1941).**
While the French combination of the Polytechnic School and the Army School dominated Portuguese engineering during the Regeneration, the Technical Institute (IST) would assume the role of main engineering institution from 1911 onwards inspired by the German model of the *technische Hochschulen*. The main protagonist of this transformation was Alfredo Bensaúde, the first director of IST, who completed his engineering training in Germany at the Technical School of Hannover, the Mining School of Clausthal, and the University of Gottingen.
© Public domain. In *Técnica*, XVI, 116, January 1941.

ing, students were now required to undergo a six-month internship in a factory.[16] Taking the direct example of German *Technische Hochschulen*, the courses combined theoretical classes and practical work at laboratories, museums, and factories, and ended with a six-month probation period in industry. In 1898, the Institute had thus definitely abandoned its initial focus on vocational training and the preparation of workers and foremen for industry, leaving that role to new local industrial schools, becoming instead a national school for higher education, providing

---

[16] Decree, June 30, 1898. *Diário do Governo*, no. 150 (1898).

specialized industrial training in chemistry, electricity, mechanics, public works, mining, and telegraphy, which combined theory, practice, and experience of the actual industrial conditions. For the first time, the students that completed the Industrial Institute course were granted the title of industrial engineers, importantly finishing with the practical monopoly of the tittle by the Army school.[17]

Bensaúde would have to wait for the new political regime to fully institutionalize his proposals. After the proclamation of the Republic in 1910 the Industrial Institute was renamed in 1911 Technical Institute in the framework of a general reform of the university system, which finally offered engineering education the same status of the training offered at the University of Coimbra. Brito Camacho, as Minister of Foment (the new denomination of the Ministry of Public Works), was the man behind the reform. The Army School was from then on (December 1910) devoted exclusively to military engineering and the civil and mining engineering courses were moved to the Technical Institute now directed by Alfredo Bensaúde. The choice of Bensaúde was not innocent. Brito Camacho wanted to introduce a radical change in technical teaching, much closer to the demands of the republican nationalist industrial agenda. The Technical Institute embodied the new ideal of a close relationship between engineers and economic development, mainly on an industrial basis, leaving behind the Portuguese engineer in the service of the state and putting an end to the old *status quo*.

The change of name deserves further reflection. Considering the centrality of industry for the entire project and that the institute didn't change premises, one might have expected that Bensaúde would have felt comfortable with the previous Industrial Institute denomination. In fact, the continuities between the Industrial Institute and the Technical Institute are too obvious to keep talking of the two as totally separate institutions as the historiography generally does by taking at face value the claims of the historical actors. Contextualizing the decision about the new name helps us understand what was at stake. The new name was a direct translation to Portuguese of *technische Hochschule* and meant to signal how the institute was responsible for bringing the German notion of *Technik* to Portugal.

As historians of technology have abundantly demonstrated, German engineers at the end of the 19th century embraced *Technik*, as a way of improving their cultural status.[18] *Technik*, understood as "the entirety of rules, procedures, and skills for achieving a specific goal," offered German engineers the ability of fashioning themselves as members of high culture, making technical education into a form of Bildung, of cultivation.[19] This broad understanding of *Technik* as an integral part of culture, instead of narrower uses referring exclusively to "the material

---

[17] The Polytechnic Academy of Oporto could also grant the tittle of engineer. In 1899, the Association of Engineers had 241 members trained in the Army School and 28 in the Oporto Academy.

[18] Adelheid Voskuhl, "Engineering philosophy: Theories of technology, German idealism, and social order in high-industrial Germany," *Technology and Culture*, 57, (2016), pp. 721–52; Eric Schatzberg, *Technology: Critical History of a Concept*, (Chicago: Chicago University press, 2018); Mikael Hård, "German regulation: The integration of modern technology into national culture," in *The Intellectual Appropriation of Technology: Discourses on Modernity, 1900–1939*, Mikael Hård and Andrew Jamison, Eds., (Cambridge, MA: MIT Press, 1998), pp. 33–67.

[19] Schatzberg, *Technology*, pp. 104–105.

Figure 6.4: **Tecnica (1926).**
The journal *Tecnica* published by the Technical Institute students confirmed these were being trained in *Technik*, the German concept that referred to technology as an integral part of culture, making Portuguese engineers into full members of a cultural elite.
© Public domain.

aspects of industry and trade," granted German engineers enough social prestige for the Kaiser to give in 1900 *technische Hochshulen* the right of awarding doctoral degrees as other universities of the Reich. The Portuguese equivalent of such recognition was the referred reform of 1911 that placed the Technical Institute at the same level of the University of Coimbra. Bensaúde was thus following the example of the *technische Hochshulen* he had studied in, institutions that trained men in *Technik*, making them members of the cultural elite. For engineers educated in *Technik*, it was thus not strange to share the pages of the magazines gathering the rest of Portuguese intelligentsia. It is also no coincidence that the journal of the student's association of the

Technical Institute was named *Técnica*, the Portuguese translation of *Technik* before technology filled up that role.[20]

The Technical Institute had a general course of two years; after this preparatory course, students could choose among five courses devoted to mining, mechanics, electricity, industrial chemistry or civil engineering. All the courses had theoretical classes and a strong practical component "essential to the technical knowledge."[21] Besides this practical training during school time, students had frequent contacts with the industrial milieu for instance by visiting factories. During the summer break, students had also a one-month probation in a factory, in order to "feel" the industrial atmosphere. This insistence in practice shouldn't distract us from the fact that the Technical Institute didn't aim at offering new skills to workers to make them foremen or industry directors as it had been the original intention of the Industrial Institute. Here we are talking instead of knowledge about practice, of *Technik* forming a new middle class of managers, between workers and owners.

## 6.3    INDUSTRY AND THE TECHNICAL INSTITUTE ENGINEERS

In addition to new curricula, the Technical Institute also introduced changes concerning faculty, in order to improve pedagogical and scientific quality according to Bensaúde's views of what constituted proper training in engineering. In 1911–1920, 12 out of the 27 professors had studied abroad at the most famous engineering schools, such as the *École des Ponts et Chaussées*, the *École des Mines*, the University of Liège, the University of Berlin, or the University of Gottingen. Most of them had previous experience in teaching in Portugal or abroad and, more importantly, a significant part of these men had a professional industrial experience. As Bensaúde insisted, "the best teachers are those who, besides teaching, have a professional experience in their own field of expertise; they know best how to manage their students' professional careers. The school is also a *bureau de placement* which allows the more talented students to find a job and economy itself to profit from these young engineers."[22]

While no more than 1% of the 300 members of the Association of Civil Engineers were employed by industry at the turn of the century, no less than 30% of the first generation of engineers trained at the Technical Institute—those finishing graduating between 1913 and 1920—were recruited by industrial companies.[23] And if we look at those graduating as chemical, mining, mechanical, or electrical engineers, thus excluding civil engineers, the numbers reach over 50% of Technical Institute graduates employed by Portuguese industry. While this seems to represent a drastic shift, one should be careful in the interpretation of such numbers. The main

[20] The translation of *Technik* as technology was due to American social scientists of the 1910s. The name *Técnica* was actually only used after 1925. Before that a first version of the journal was called Técnica Industrial.

[21] Bensaúde, *Notas*, p. 39.

[22] Bensaúde, *Notas*, pp. 68–69.

[23] Nuno Madureira, "Visionários e dirigentes: os engenheiros portugueses na primeira metade do século XX," *Actas do XX Encontro da Associação Portuguesa de História Económica e Social* (Porto: APHES, 2000).

justification is the simple fact that students previously graduating from the Industrial Institute had no right to the formal tittle of engineer. As mentioned, only after 1898 did the students graduating from the Industrial Institute get the tittle of engineer. Until then, this was reserved for those completing the engineering course of the Army School, in addition to the few trained at the Polytechnic Academy of Oporto. But it shouldn't be overlooked the fact that the technical training at the Industrial Institute, which previously had held a lower status than that granted at the Army School, was now considered to be at the forefront of engineering in the revamped institution.

It is also important to understand that the Technical Institute engineers were not evenly distributed among the country's different industries. At the beginning of the 20th century, textile factories and flour milling constituted the major industrial sectors in Portugal in terms of capital invested and horsepower, nevertheless they didn't employ a single graduate from the Technical Institute. The main employers of its students were instead the conglomerate of *Companhia União Fabril* (CUF), famous namely for its chemical fertilizers produced from Portuguese pyrites and that would become in the following years the main industrial company in the country; the *Companhia União Metalúrgica*, the single steel mill in the country equipped with two Bessemer converters; the Burnay investment bank, a major player in the hydroelectric projects that would flourish in the following years; the Uranium mines of Urgeiriça; the *Companhia Portuguesa de Fósforos*, holding the monopoly of matches production in Portugal. These companies represented sectors—chemical, metallurgical, energy, mining—that would gain increased importance in the upcoming years and importantly change the nature of Portuguese industrialization.

The history of the Technical Institute chemistry professor Charles Lepierre is eloquent about the entanglements between academic engineering and industry weaved by the new institution. Bensaúde hired Lepierre to the institute right after the 1911 reform, following his strategy of relying on foreign professors with experience in Portuguese higher education: after graduating in 1887 from the Parisian *École de Physique et Chimie Industrielles*, Lepierre would move to Lisbon the following year to teach chemistry both at the Polytechnic School and at the Industrial Institute. He didn't last long in the post, and one year after he was moving to Coimbra, first teaching at the local Industrial School *Brotero* and later at the University where he would become the chair of sanitary engineering. Lured by Bensaúde's offer to return to the reformed engineering Institute, he would join the Technical Institute faculty in 1911 and demonstrate in the following 26 years the importance of chemical analysis, organic chemistry, and technological chemistry to establish the new role of the institution in the national economy. His chemistry laboratory at the institute asserted the good quality of numerous Portuguese spring waters and promoted spa tourism all over the country; contributed to the development of canned fish industry; analyzed sands from the Gândara dunes enabling the building of a major cement factory—*Maceira-Lis*—in the most inhospitable region of Portugal. In 1915, Lepierre would become manager of the *Fábrica dos Produtos Químicos da Póvoa de Santa Iria*, dedicated mainly to the production of fertilizers, and two years after he would start consulting work for the *Com-*

Figure 6.5: **Chemical Engineer Charles Lepierre (first from the right) in the Uranium Mines of Barracão.**
Charles Lepierre represented the new model of the Technical Institute professor, with foreign training and direct connections to industrial undertakings dependent of national resources such as Uranium mining.
© Arquivo Histórico do IST.

*panhia Portuguesa de Fósforos*. Even more importantly, in 1917 he became consultant for the Burnay investment bank, a relation that would endure for no less than 20 years.

The bank had hired Lepierre to direct the works of its uranium ore treatment plant in *Barracão* in the remote hilly region of *Beira Alta*, among the poorest parts of the Portuguese territory, isolated from the main urban and industrial centers.[24] Lepierre had been student of no less than Pierre Curie and his chemical expertise would reveal invaluable in making Portugal, even if only for a brief period of time, a main supplier of Radium to the Curies profitable technoscientific endeavor that connected physics, industrial operations, and medical applications.[25] After the Austrian embargo of 1907, which halted the export of Uranium ores from the Joachimsthal mines in Bohemia, the granite interior region of Portugal became key to the Curies. And while the ores were initially treated in France, by 1918 Lepierre boasted that his

---

[24] Ricardo Moreira, "O arquivo do cancro. Um estudo da cultura material e tecnológica na génese da medicina oncológica em Portugal, (1912–1926)," MSc dissertation, Instituto de Ciências Sociais, Universidade de Lisboa, 2013.

[25] Xavier Roqué, "Marie Curie and the radium industry: A preliminary sketch." *History and Technology*, 13.4, (1997), pp. 267–291; Soraya Boudia, "The Curie laboratory: Radioactivity and metrology." *History and Technology*, 13.4, (1997), pp. 249–265.

Portuguese plant produced 3 g of RaBr2, making it the first source of radium in the world.[26] To produce 10 mg of RaBr2, one had to use no less than 3 tons of expensive reagents, which demanded the optimization of the process. Only a thorough chemical analysis of the ores could determine the quality and quantity of reagents, which until the arrival of Lepierre to Barracão were used with no criteria, jeopardizing the whole operation. In the following years, Technical Institute graduates would become the main technicians of the plant of the Burnay controlled Société Urane-Radium. By 1929 there were some 99 uranium concessions in Portugal, connecting the most remote areas of the country with one of the most sophisticated industrial undertakings in the world. The discovery of rich ores in the Belgian Congo and the monopoly of the market by the *Union Minière* from the mid-1920s onward until the World War II would bring to an end this first Uranium rush in Portugal.

The conversion by chemical and mining engineering of vast areas of the backward mountainous region of Beira Alta—where communal lands and transhumance practices were still common—into object of financial speculation as uranium concessions supplying the Curies laboratory in Paris, has certain futurist undertones. And thus, while Campos published his afforestation and hydroelectric proposals next to the texts of the republican intellectual elite of *New Harvest* led by António Sérgio, one shouldn't be surprised to find engineers involved in even more radical plans that made them legitimate members of the cultural avant-garde. Geraldo Coelho de Jesus, the mining engineer responsible for managing the coal mines of Porto de Mós, counted with no less than with the enthusiastic involvement of the modernist poet Fernando Pessoa to publicize his "Basis for an Industrial Plan" of 1919.[27] The plan didn't differ too much from the one by Campos in its combination of technical training, tariff protection, credit, and, of course, hydroelectricity as magical key to industrialize the backward nation. But the ideology of technical elites guiding the country promoted by Jesus went further than Campos, claiming not only for a more direct intervention of the State in the industrialization effort, but also for more radical authoritarian politics to undertake the ambitious reengineering of Portugal.

The poet and the engineer, Pessoa and Jesus, formed a movement—the Nucleus for National Action (*Núcleo de Acção Nacional*)—demanding the combination of technical expertise with charismatic political leadership referring the example of the short-lived dictatorship of Sidónio Pais (December 1917–December 1918). The dictatorial rule had promised to end the social turmoil produced by World War I shortages, namely the strikes and food riots of which the most notorious had been the "potato revolution" of May 1917 when the Lisbon populace

---

[26] Charles Lepierre, "Radioactividade. Conferência realizada na associação dos engenheiros civis Portugueses em junho de 1924." Arquivo do IST, Fundo Charles Lepierre, PT/IST-AP-PCL; Charles Lepierre, "Contribution a l'étude des minerais d'uranium-radium portugais," *Revista de Química Pura e Aplicada*, (1933), pp. 15–26; Ricardo Moreira, "Arquivo do cancro."

[27] Manuel Villaverde Cabral, "A estética do nacionalismo: Modernismo literário e autoritarismo político em Portugal no início do século XX," *Novos Estudos CEBRAP*, 98, (2014), pp. 95–122; José Barreto, "O núcleo de acção nacional em dois escritos desconhecidos de Fernando Pessoa." *Pessoa Plural: Revista de Estudos Pessoanos/Journal of Fernando Pessoa Studies*, 3, (2013), pp. 97–112.

looted groceries and warehouses leaving behind 38 deaths, 117 wounded, and more than 500 arrests.[28]

It has been importantly noted by political historians that Sidónio's brief rule inaugurated the experiments with charismatic authoritarian dictatorships of the interwar period in Europe.[29] For our argument, it is no less important to remember that Sidónio Pais was trained at the Army School (the main engineering school until the transformation of the Industrial Institute into Technical Institute), that he had directed the Industrial School Brotero in Coimbra—the same where Charles Lepierre had also lectured—and that he was a professor of mathematics at the University of the same city, responsible for no less than the chair of differential and integral calculus, the main mathematical tool of the modern engineer to intervene in the world. In 1910 he was a member of the administration of the Portuguese Railways Company and in 1911 he had a first brief experience in government as Minister of Foment (Public Works). This political involvement came through his friendship with Brito Camacho, the previous minister that issued the decree founding the Technical Institute in 1911, and leader of the right wing of the Republican movement. Critical of the populist tendencies of the Republican Party—the democrats—Camacho assembled the Republican economic and technical elites in a new party— the unionists—in which Sidónio Pais participated.[30]

To sum up, and for the reader to not get lost in the confusing political scene of the Portuguese years of the Republican regime (1910–1926), what we are suggesting is the dictator Sidónio embodied an original combination of charismatic political leadership with technical elitism, advancing top-down solutions to bring order and development to allegedly decaying Portugal. The deep sense of crisis that started with the British *Ultimatum* of 1890 was only intensified by the difficulties of the World War I, opening up the field for more radical political options in which engineers were prominent.

Fernando Pessoa, the most inspired of the modernists, and deservedly celebrated as the most important Portuguese poet of the 20th century, published in 1917 in the avant-garde journal Futurist Portugal (*Portugal Futurista*) under the name of Álvaro de Campos—his heteronym[31] trained in mechanical and naval engineering in Glasgow—his violent and controversial poem *Ultimatum*, non-veiled reference to the British ultimatum. After announcing his "Evic-

---

[28] Vasco Pulido Valente, "A revolta dos abastecimentos: Lisboa, maio de 1917" in *Tentar Perceber*, (Lisboa: INCM, 1983), pp. 183–197.

[29] Manuel Villaverde Cabral, "A grande guerra e o sidonismo (esboço interpretativo)." *Análise Social*, (1979), pp. 373–392; Maria Alice Samara, "Sidonismo e restauração da república. Uma 'encruzilhada de paixões contraditórias,'", in Fernando Rosas e Maria Fernanda Rollo, org., *História da Primeira República Portuguesa*, (Lisboa: Tinta da China, 2010), pp. 371–395.

[30] Significantly enough, Sidónio's coup of December 1917 was engendered in the offices of the newspaper *A Luta*, (The Fight)—the mouthpiece of the unionists—and would count with the support of several unionist ministers and important industrialists, namely Alfredo da Silva, the leader of the CUF conglomerate.

[31] Fernando Pessoa is particularly famous for his heteronyms under whose name he wrote some of his best poetry. Pessoa's heteronyms are much more than mere pseudonyms, since he gave them full-fledged personalities and biographies. The most famous of Pessoa's heteronyms are Álvaro de Campos, Alberto Caeiro and Ricardo Reis. Jerónimo Pizarro and Patricio Ferrari counted no less than 136 different authors created by Pessoa, among heteronyms, protoheteronyms, and pseudonyms: Patricio Ferrari and Jerónimo Pizarro Eds., *Eu Sou Uma Antologia. 136 Autores Fictícios*, (Lisboa: Tinta da China, 2016).

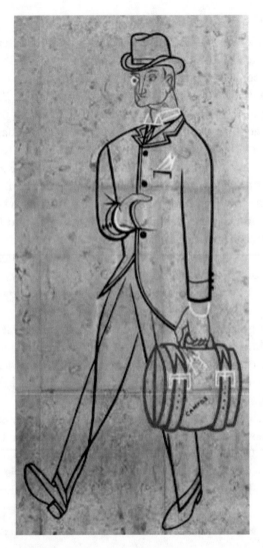

Figure 6.6: **University of Lisbon mural depicting the futurist poet-engineer Álvaro de Campos by Almada Negreiros.**
Fernando Pessoa, the most important Portuguese poet of the 20th century, published under the name Álvaro de Campos his most celebrated futurist poems. Campos, trained in mechanical and naval engineering in Glasgow, announced in 1917 the coming of a new era of poet-engineers. © Authors' photo.

tion notice to the mandarins of Europe"—the European cultural and political European elite that led to the carnage of the War, thus issuing his own *Ultimatum* to Anatole France, Kipling, Yeats, and Bernard Shaw as well as to Wilhelm II, David Lloyd George, and Venizelos—Pessoa (Campos) loudly announced a new era:[32]

> "…I proclaim the inevitable coming of a Humanity of Engineers!
>
> More than that, I absolutely guarantee the coming of a Humanity of Engineers!
>
> I proclaim the imminent scientific creation of Supermen!
>
> I proclaim the coming of a perfect, mathematical Humanity!"

For Pessoa, the embodiment of the Nietzschean superman, the new demiurgic poet, was an engineer. Pessoa would also publish a much less inspired ode to Sidónio Pais—the "President-King" as he called him—after his murder in 1918. But the important thing to retain here is how engineering, namely as being practiced at the Technical Institute, was presented as a practice belonging to the same sphere of avant-garde culture and radical politics: engineers (Jesus) were as modernist as poets (Pessoa) in reinventing Portugal. In fact, the most celebrated of Pessoa's creations—his heteronym Álvaro de Campos—was a combination of Pessoa and Jesus, a poet-engineer, with his *Ultimatum* announcing the new era of poet-engineers.

Not every poet was as modernist as Pessoa and not every engineer was a declared supporter of Sidónio as Jesus. Nonetheless, the political instability that followed the World War I and the return of political control by the populist Democratic Party after the murder of Sidónio would progressively make engineers more modernist, easily embracing more politically radical right wing alternatives to democracy. Campos' trajectory is again illuminating. While he had welcomed the republican coup of 1910 convinced that the new democratic regime would reform the country according to his conservationist proposals, in 1923 Campos overtly embraced authoritarianism to "heal Portugal." Making no case of the republican constitution, he pleaded for a "national ministry of public salvation…invested with exceptional powers…to launch the bases of national reorganization."[33]

## 6.4    THE DICTATORSHIP AND THE RATIONAL MANAGEMENT OF THE COUNTRY'S RESOURCES

So it is no surprise to find the name of Campos as minister of commerce and communications (the new denomination of Foment) of the first government formed after the military coup d'état of May 1926 that overthrew the Republic and inaugurated the authoritarian rule that would last until no less than April 1974. Considering that the political situation was still too uncertain,

[32] Fernando Pessoa, *Sensacionismo e Outros Ismos*, Edição crítica de Fernando Pessoa, vol. X, editado por Jerónimo Pizarro, (Lisboa: INCM, 2009), pp. 232–278.

[33] Ezequiel de Campos, *A Crise Portuguesa: Subsídios para a Política de Reorganização Nacional, em Ezequiel de Campos, Textos de Economia e Política Agrária e Industrial, 1918–1944*, Ed., Fernando Rosas (Lisboa: Banco de Portugal, 1999), p. 175.

Campos didn't accept the nomination, but his proposals we detailed above were soon made into law. Only five months after the coup, a new law for hydraulic concessions was issued (*lei dos aproveitamentos hidráulicos*), with the state funding and regulating both the production (new dams) and transmission (high voltage) of electricity. This law would constitute the basis for much of the dam building craze of the following years. A dictatorship thus seemed to many engineers the fastest and the only feasible way to accomplish the ambitious task of reorganizing Portugal through a rational management of the country's resources, the modernist project of engineers.[34] More than supporting the dictatorship, engineers helped produce it, offering it much of the projects for reorganizing national life.

And if Campos' case is revealing enough, the one of Duarte Pacheco (1900–1943) as director of the Technical Institute (1926–1943) and flamboyant minister of Public Works (1932–1936 and 1938–1943) of the authoritarian New State is even more significant. His role as director of the most important engineering institution of the country is directly related with the vexing issue of the Technical Institute facilities during the first decades of the 20th century.[35] How could engineers incarnate a modernist ethos while still speaking from the 19th-century undignified warehouses of the Industrial Institute in the Boavista neighborhood? Bensaúde reformulated curricula, hired new professors, and established new ties with emergent industrial sectors of the Portuguese economy. But he had clearly failed in gaining approval for new facilities emulating the *Technische Hochshulen* he had in mind when making those reforms. In 1927, when Pacheco became its director, the Technical Institute still resembled more the now outmoded Parisian *Conservatoire des Arts et Métiers* than the expansive *Technische Hochshule Charlottenburg* in Berlin.

A plausible explanation for Duarte Pacheco becoming director of the Technical Institute at 27 years old is his political connections. Already his nomination as professor in 1925 countered the institutional policy defined by Bensaúde of hiring only faculty experienced in both the academic and industrial worlds. Duarte Pacheco had in fact no experience outside academia, having finished his degree in electrical engineering at IST in 1923. But the argument is not one of political patronage overcoming meritocracy, but one of using politics to advance the interests of the institution. It's not that politics influenced the institution to nominate Duarte Pacheco, but that he was nominated to influence politics. Besides belonging, of course, to the right wing elitist faction of Republicanism,[36] he had close connections in the Lisbon municipality and in the Lisbon finance milieu that would reveal strategic in providing the Institute with the large laboratories and workshops dreamt by Bensaúde. His easy access to the municipal power was granted by his close friendship with the alderman Caetano Beirão da Veiga—professor of accounting at the institute and Pacheco's mentor—which enabled the real estate operation behind

---

[34] Campos, *Pregação*, pp. 136–138.

[35] Sandra Vaz Costa Marques de Almeida, "O país a régua e esquadro: Urbanismo, arquitectura e memória na obra pública de Duarte Pacheco." PhD dissertation, Universidade de Lisboa, 2009.

[36] Having joined the Liberal Union of Cunha Leal, the new name of the unionists of Brito Camacho in coalition with the evolutionists of António José de Almeida. This political faction, defended the government of technical elites criticizing the populism of the Democratic party.

the funding of the new building. The Technical Institute acquired a vast rural lot in the city outskirts in the northeast that was soon transformed by the municipality into urban area. The area was much larger than the one the institute needed and much of it could be now sold with great profit as urban property. This typical real estate speculation, profiting from influence in city hall to transform rural land into urban property, would constitute the basis for financing the construction of the new facilities. Pacheco's brother—a lawyer working in the insurance business—would put him in connection with the bank *Caixa Geral de Depósitos* for financing the whole operation. After guaranteeing the contribution of the young professor of architecture of the technical Institute Porfirio Pardal Monteiro to design the new building, construction started in 1929.

While professors in the Boavista complained of passing by tramways altering electrical measurements in the Physics laboratory, the lack of natural light for drawing classes, the shabbiness of amphitheaters, or the insalubriousness of the neighborhood, when confronted with the scale of the new Technical Institute proposed by Pardal Monteiro they felt vertigo. Some expressed their doubts about the ability of the institution to fill up the generous spaces designed by the architect. But Bensaúde had no doubts this was the appropriated scale needed for the practice of modern engineering, offering his continuous support to Pacheco's undertaking. Pardal Monteiro's architectural design granted each engineering discipline one independent modular building, flexible enough to grow in function of the discipline's internal development. Distributed in an elevated terrain in U shape, the central building concentrated the direction and administrative offices of the institute, the library, the main amphitheaters, and the civil engineering department. The sides housed the four remaining departments—mechanical, electrical, chemical, mining—in addition to workshops and social amenities such as a gym and a swimming pool. The identity of the engineer was not only to be shaped in classrooms and laboratories but also through the cultivation of a healthy body.

Only the main building had a few Art Deco concessions that allegedly enhanced the monumentality of the engineering school. Otherwise, the straight lines, clean façade, large unframed windows, horizontal roofs, modular buildings, and use of concrete made the IST the largest scale modernist structure in Lisbon. Located at the top of the new Afonso Henriques parkway, the Technical Institute campus had the effect of an acropolis indicating the direction of Lisbon's urban expansion. Adjacent to it, the new mint and the National Institute of Statistics, designed in the same style also by the architect Pardal Monteiro and with the contribution of the civil engineer José Belard da Fonseca, would shortly follow.[37] Bordering the institute's northern fence, the social housing project of *Arco do Cego* would also change its aims. Born as a petty project of the Republic to solve the housing problems of Lisbon working-class families, it was redesigned to house instead the municipality own workers and, more importantly to this text, to include upscale chalets for Technical Institute engineering professors. The most celebrated in

---

[37] João Pardal Monteiro, "Para o projeto global: Nove décadas de obra. Arte, design e técnica na obra do atelier Pardal Monteiro," PhD dissertation, Universidade Técnica de Lisboa, 2012.

Figure 6.7: **The campus of the Technical Institute inaugurated in 1936.**
The straight lines, the clean façades, the large unframed windows, the horizontal roofs, the modular buildings, and the use of concrete made the Technical Institute the largest modernist structure in Lisbon.
© Photo by Pinheiro Correa. Arquivo Fotografico da CML.

the canons of Portuguese modernist architecture was the one that Cristino da Silva designed for the structural engineer José Belard da Fonseca, who would later be director of the Institute from 1942–1958. While Belard da Fonseca made the calculations for the innovative concrete structure of the *Capitólio* Theater that would earn Cristino da Silva the status of modernist architect in the annals of the discipline; the architect designed the geometrical façade, modular volumes, and streamlined furniture adequate to house the modernist engineer.

The Lisbon mayor and general José Vicente de Freitas, who together with Duarte Pacheco reimagined this Arco do Cego neighborhood as modernist showcase dominated by the imposing building of the Technical Institute, would soon become prime minister of one of the first governments coming out of the 1926 military coup d'état. Freitas invited Duarte Pacheco to become his minister of education asking him as well to convince the professor of the University of Coimbra Oliveira Salazar to assume the position of finance minister. Salazar accepted Pacheco's arguments. Pacheco lasted less than eight months as minister leaving the government

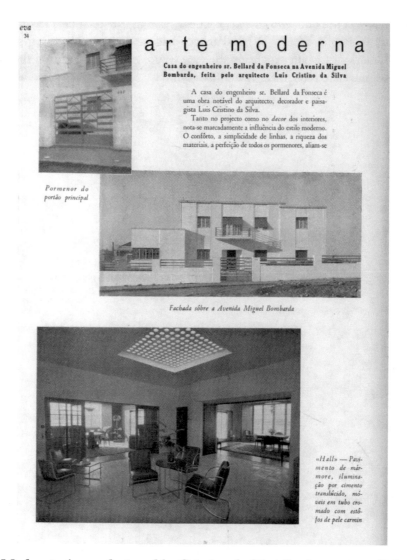

Figure 6.8: **Modernist house designed by Cristino da Silva for the engineer Belard da Fonseca.**

While the Technical Institute engineer Belard da Fonseca made the calculations for the innovative concrete structure of the Capitólio Theater that would earn Luís Cristino da Silva the status of modernist architect in the annals of the discipline; the architect designed the geometrical façade, modular volumes, and streamlined furniture adequate to house the modernist engineer. © *Eva: jornal da mulher e do lar*, 1932.

in November 1928. But in 1932, Salazar, already in his condition of uncontested leader of the fascist New State, was the one who invited Pacheco, asking the Technical Institute director to become Minister of Public Works. While Pacheco had made his engineering school into the main urban landmark of Lisbon expansion, Salazar would now offer Pacheco the possibility of mobilizing its engineers to reengineer the entire country.

# CHAPTER 7

# Engineering the Fascist New State

Engineers would become crucial players of the New State (Estado Novo), the authoritarian fascist regime that ruled the country from 1933–1974.[1] The New State replaced liberal mechanisms of representation by ideological nationalism, the one-party state, systematic political and social repression, and social and economic corporatism.[2] For corporatist ideologues, nuclear families and corporations of producers were the constitutive organs of society, not the individuals of liberal theories or, even worse, the classes of Marxism. A "Strong State"—one of the propaganda mottos of the regime—eliminated political dissent, sponsored the cartelization of economic sectors to avoid the effects of unrestricted capitalism, and imposed "cooperation" between workers and industrialists through state controlled syndicates while violently repressing workers' unions.

The engineer José Araújo Correia, minister of commerce and communications of that dictatorial government in which Duarte Pacheco and Salazar first coincided, insisted in a speech in 1934 at the first convention of the National Union—the single party of the regime—on the notion of fascism as third way, an alternative to both liberal democracies and communist Russia: "Fascism in Italy, Hitlerism in Germany, the evolving pacific revolution in Portugal are all spontaneous reactions stemming from people's wishes to enlarge to under protected classes the benefits of economic progress. Neither democracies nor socialism protected proletarians."[3]

The empowering of the state to actively intervene in the coordination of the whole national economy and society would prove fertile ground for the employment of engineers. The prevailing image of Portuguese fascism, dominated by a traditionalist establishment and reduced to the

---

[1] On the fascist nature of the regime see Manuel de Lucena, "Corporatisme au Portugal, 1933–1974. Essai sur la nature et l'ambiguité du regime salazariste," in Didier Musiedlak, Ed., *Les Expériences Corporatives dans l'Aire Latine*, (Bern: Peter Lang, 2010), pp. 153–202; Luís Reis Torgal, *Estados Novos Estado Novo*, (Coimbra: Imprensa da Universidade de Coimbra, 2009); Fernando Rosas, *O Estado Novo nos Anos Trinta, 1928–1938*, (Lisbon: Estampa, 1996); Jorge Pais de Sousa, *O Fascismo Catedrático de Salazar*, (Coimbra: Coimbra University Press, 2011); Manuel Loff, *O Nosso Século é Fascista!": O Mundo Visto por Salazar e Franco, 1936–1945*, (Oporto: Campo Das Letras, 2008). For an alternative interpretation stressing authoritarianism and downplaying fascism, see António Costa Pinto, *Salazar's Dictatorship and European Fascism: Problems of Interpretation*, (New York: SSM/Columbia University Press, 2005); Manuel Braga da Cruz, *O Estado Novo e a Igreja Católica*, (Lisbon: Bizâncio, 1998). On continuities and discontinuities between different fascist regimes see the informed discussion in Aristotle Kallis, "Fascism, para-fascism and fascistization," *European History Quaterly*, 33, (2003), pp. 219–49.

[2] This list is taken from Manuel Villaverde Cabral, "Portuguese fascism in comparative perspective," working paper presented at the *International Political Science Association, XIIth World Congress*, Rio de Janeiro, 1982. This combination made the Portuguese New State a full legitimate member of the European family of fascist regimes.

[3] José Araújo Correia, *Directrizes Económicas do Estado Novo*, (Lisboa: Sociedade Nacional de Tipografia, 1934), p. 16. On this engineer see, Carlos Bastien, "O economista Araújo Correia." *Estudos de Economia*, 3, (1985), pp. 291–320.

Figure 7.1: **Portuguese pavillion at the Paris World Exhibition, 1937.**
The prevailing image of Portuguese fascism, dominated by a traditionalist establishment and reduced to the trinity trinity "God, Fatherland and Family," doesn't do justice to the relevance of the technological dimension in the making and growing of the New State.
© Biblioteca de Arte da Fundacao Calouste Gulbenkian.

trinity "God, Fatherland and Family," doesn't do justice to the relevance of technical elites in the making and growing of the New State.[4]

## 7.1   REGAINING POWER: THE "SOCIAL ROLE" OF PORTUGUESE ENGINEERS IN THE DICTATORSHIP

Engineers were not passive historical actors mobilized by the new authoritarian regime. They actively made a space for themselves in the new political situation. The organization of the 1st National Meeting of Engineering in 1931 speaks volumes concerning this self-mobilization of engineers for the New State. Three hundred and fifty-seven engineers (all male) registered for

---

[4] For a fuller exploration of the relation between technical and scientific elites and the making of fascist regimes in Portugal, Italy, and Germany; see Tiago Saraiva, *Fascist Pigs: Technoscientific Organisms and the History of Fascism*, (Cambridge, MA: MIT Press, 2016).

an event aimed at asserting their role in reorganizing national life on technical principles. The Association of Civil Engineers, responsible for the gathering, presented it as a unique opportunity to discuss technical topics in the context of the Portuguese modernizing agenda; to show the vitality and work of Portuguese engineers; and to promote the cult of engineering among the public.[5]

More than an exercise in professional introspection, this was a major propaganda effort materialized in the ambitious exhibition organized for the meeting. Aerial photographs, scale models, and film, took over the Palace of Fine Arts in Lisbon announcing to the public the new technological Portugal crossed by roads, high voltage cables, and telephone lines. The choice of Cottinelli Telmo as the architect responsible for the design of the exhibition couldn't be more telling. Telmo would become in the following years the producer of major propaganda events such as exhibitions, festivals, or movies that established the public image of the fascist New State.[6] He creatively combined modern technological elements with nationalistic references to a mythical glorious past of knights and sailors, in what would become the distinctive imagery of the regime, or what an inspired scholar called "futurism of the past."[7]

The 1st National Engineering Meeting had a clear popularization agenda: Portuguese engineers wanted to "show to the general public their work."[8] The agenda was twofold: on one hand the exhibition displayed a large number of engineering works and engineering instruments, dazzling the audience with the power of technology; on the other hand, the papers and lectures presented at the Meeting, the visits to industrial facilities and public works, and the official speeches, all of them extensively described in both national and regional newspapers, asserted Portuguese engineers as key players in remaking the country.

It would be quite difficult for any reader to ignore the event and escape its powerful vision of a technological Portugal. The "gee-whiz effect" produced by the exhibition was enhanced both by the amount of technical details presented in the Lisbon and Oporto newspapers (for instance, the painstaking descriptions of the contents of conference papers and of exhibition stands) and by wrapping it all up in a hymn of praise to engineering. The presence of the President of the Republic and the Minister of Commerce at the opening and closing ceremonies of the meeting confirmed the political alignment of the engineering profession with the dictatorship.

Newspapers suggested less grandiose more mundane ways of consuming the event. They detailed the presence of elegant and cheerful ladies among the technological artifacts and invited visitors for having tea and taking a dive at the swimming-pool in fashionable Estoril or for drinking beer and eating sandwiches among workers at the Alfeite shipyards. The lyric rhymes at the closing banquet of the conference by the "Camões of Engineering" (a reference to the

---

[5] Maria Paula Diogo and Ana Cardoso de Matos, "Going public: The first Portuguese national engineering meeting and the popularization of the image of the engineer as an artisan of progress, (Portugal, 1931)." *Engineering Studies*, 4, no. 3, (2012), pp. 185–204.

[6] Margarida Acciaiuoli, *Exposições do Estado Novo, 1934–1940*, (Lisbon: Livros Horizonte, 1998).

[7] Herminio Martins, *Classe, Satus e Poder e Outros Ensaios Sobre o Portugal Contemporaneo*, (Lisboa: Imprensa de Ciências Sociais, 1998).

[8] *Diário de Notícias*, 23474, (June 8, 1931), p. 1.

Figure 7.2: **1st National Meeting of Engineering, 1931.**
Three hundred and fifty seven men participated in the conference organized by the Association of Civil Engineers asserting the self-mobilization of engineers for the fascist New State. The poster was designed by Cottinelli Telmo, a leading figure of the New State propaganda agenda. © Public domain.

national poet of the 17th century), also aimed at establishing a connection between engineers and the general public.

Although there were few explicit references to the growing technocratic tide that had reached both Europe and the United States by the early 1930s, Portuguese engineers were certainly aware of it. By using the 1st National Engineering Meeting to create an atmosphere of technology-driven euphoria, showing technology as moral force and national imperative, Portuguese engineers positioned themselves as decision-makers of the new authoritarian regime. At the end of the meeting, they had good reasons to feel they had successfully made their case.

Duarte Pacheco, in his double condition as director of the Technical Institute and minister of Public Works from 1932–1936 and from 1938–1943 (the year of his death), embodied this identification of engineering with the building of the new regime. The most charismatic of Salazar's ministers directly intervened in the design of new roads, city plans, ports and air-

Figure 7.3: **The engineer Duarte Pacheco (left) with the dictator Oliveira Salazar (right) inaugurating the Exhibition of the Portuguese World in 1940. Between the two, the minister of education (Carneiro Pacheco) and the minister of commerce and industry (Costa Leite). Further back (left), Antonio Ferro, the head of propaganda of the regime.**
Duarte Pacheco, in his double condition as director of the Technical Institute and minister of Public Works from 1932–1936 and from 1938–1943 (the year of his death), embodied the identification of engineering with the building of the fascist regime. The most charismatic of Salazar's ministers directly intervened in the design of new roads, city plans, ports and airports, urban parks, monuments, or new stadiums.
© Biblioteca de Arte da Fundacao Calouste Gulbenkian.

ports, urban parks, monuments, or new stadiums. What was first tried in the exhibition of 1931 became an obligatory element of every celebration of the regime: technological infrastructures were presented side by side with the picturesque features of the Portuguese landscape, the glorious feats of the Portuguese nation, and the social peace brought by the dictatorship. In other words, Pacheco was the New State first "supplier of matter with which to make propaganda of itself."[9]

The program for the building of affordable houses launched in 1933, the year of the institutionalization of the regime by a new constitution, is an excellent example of how engineering contributed to publicize and materialize New State ideology.[10] Despite the many promises of the previous democratic regime to solve the housing problems of working classes, none of its projects had been concluded. To prove the New State ability to succeed where the Republic had failed, Duarte Pacheco created a new section in his Ministry of Public Works dedicated to the design of affordable housing, which in the long term would lead to the building of some 16,000 houses. The above-mentioned engineer José Araújo Correia, commissioned by the Ministry, traveled in 1934 to Germany, Austria, and Hungary, all members of the fascist family, to get acquainted with international experiences in affordable housing. This was not a topic in which Portuguese engineers could claim much previous expertise, but Correia was glad to extend the realm of engineering to explicitly social problems. After all, he had published in 1927 an influential article in *Técnica*, the journal of the Technical institute, on "The social role of the engineer," in which he didn't shy away from sustaining that "engineering... solves most of the world problems."[11]

Affordable houses designs coming out of the Ministry of Public Works revealed engineers' obsession with standardization of construction to keep costs at a minimum. But significantly enough, the engineer responsible for the whole program in the Ministry decided to hire the architect Raúl Lino, famous for his theorization of the "Portuguese house" aiming at distilling an identifiable national style of architecture. While Lino didn't spare efforts at "reduc[ing] the areas to a comfortable minimum," he also acknowledged that "everything possible was done to disguise the indispensable standardization."[12] He added to the affordable houses alleged traditional elements, like pitched roofs and small porches, to distract dwellers from "the mechanical processes and Taylorism" of their workplace.[13] The affordable houses became the product of international building standards and interpretations of national style, engineering, and nationalism.

---

[9] Margarida Acciaiuoli, *António Ferro: A Vertigem da Palavra: Retórica, Política e Propaganda no Estado Novo*, (Lisboa: Editorial Bizâncio, 2013), p. 250.

[10] The description of the project is based on Sérgio Dias da Silva and Rui Jorge Garcia Ramos, "Housing, nationalism, and social control: The first years of the Portuguese estado novo's affordable houses programme." In *Southern Modernisms: From A to Z and Back Again*, Ed., Joana Cunha Leal, Maria Helena Maia, and Begoña Farré Torras, (Porto: CEAA/IHA, 2015), pp. 255–274.

[11] José Araújo Correia, "A função social do engenheiro," *Técnica*, 6, (1927), pp. 1–3.

[12] Quoted in Silva and Ramos, "Housing, nationalism and social control," p. 271.

[13] Silva and Ramos, "Housing."

Figure 7.4: **Affordable Houses, Porto**.
The nationalistic dimension expressed through architecture style was not the only fascist element of the program for affordable houses. These were designed not as collectivist complexes but as independent houses with a vegetable garden, units for the reproduction of the nuclear family, organic basis of the national community as understood by the ideologues of the fascist regime. © Centro Português de Fotografia.

The nationalistic dimension expressed through architecture style was not the only fascist element of the program for affordable houses. In fact, these were designed not as collectivist complexes but as independent houses with a vegetable garden, units for the reproduction of the nuclear family, organic basis of the national community as understood by the ideologues of the regime. Even when built in urban areas, the housing models were supposed to promote traditional family values—man as purveyor and family head; woman as dedicated spouse and caring mother; child respectful of multiple forms of authority (parents, religion, school)—associated with rural life, replicating in the city idyllic timeless rural villages. The vegetable garden not only helped the family budget and its economic independence, but should also "stop the waste of free time in places of pernicious activities."[14] Although the houses were initially planned for working class families, observance of "proper" morals and political alignment became decisive factors to select deserving tenants among the population.

Job stability was the other crucial factor, and only workers belonging to the new syndicates organized by the corporatist state were entitled to participate in the program. The Ministry of Public Works made the projects but it was the corporatist structure of the New State that decided which employees had right to affordable housing. The corporatist organisms that dominated large sectors of the Portuguese economy—federations of producers, regulatory commissions, national unions (controlled by the state)—were indeed responsible for funding half of each individual project, the other half financed by the general state budget. Importantly enough, engineers were not only present in the building dimension of the program but they also had a say in its financing. It was also an engineer, the now familiar José Araújo Correia, who as member of the board of directors of the state-owned bank, *Caixa Geral de Depósitos, Crédito e Previdência*, directed the financial operations for granting affordable house loans. The program was perhaps the best example of Correia's ideas about the "social role of the engineer" and of his insistence on the importance of economics for the fulfillment of such role. Engineers had indeed proved their value in putting in place a major social project that weaved standardized construction, nationalist aesthetics, traditional family values, and economic and social corporatism. The affordable housing program launched by Duarte Pacheco, minister and director of the Technical Institute, can be understood as a daily lecture in fascist live, with tenants under the constant menace of eviction in case of political, social, or moral dissidence.

## 7.2    IRRIGATING THE PORTUGUESE GARDEN

Salazar, the dictator, sustained that for the State to guide national economy, "the constitution should provide the building of great works such as communications, sources of power, transportation networks and electrical energy distribution…whose plans ought to be designed and enforced by the State."[15] The state not only coordinated the different economic interests, as it provided as well the technological infrastructure on which to build the "new corporatist economy." The main funds of the Law for Economical Reconstitution of 1935 went to roads, harbors, irrigation dams, public buildings, and defense expenditures, a typical depression era array of public investments.[16] The dictator's sermon offers little doubt about the ambitions of the regime: "The dominant thought in the Administration is to do nothing without a plan."[17] That same year, 1935, a Plan for Agriculture Hydraulics was set in motion; in 1936 the government launched the Commission for National Electrification and the Commission for Internal Colonization; in 1938, the Forestation Plan promised to transform large tracts of communal lands

---

[14] Silva and Ramos, "Housing," p. 272.
[15] Fernando Rosas, *O Estado Novo*, p. 58.
[16] Silva and Ramos, "Housing," p. 266.
[17] 15 anos de obras públicas, livro de ouro, p. 122, citação de Trigo de Morais.

into productive properties sustaining a new national industry. Transportation, irrigation, electricity, and forests: the modernist dreams of Portuguese engineers such as Ezequiel de Campos were becoming reality through the institutionalization of the New State. In 1936, the Technical Institute new sprawling facilities were finally inaugurated.

To explore the reach of the engineers' actions during Salazar's dictatorship, we will now detail their involvement in irrigation projects, one of the infrastructural showcases of the regime. Engineers had been able to occupy a major place in the propaganda of the New State and their college was the most distinguished building of the capital of the country, but, as we shall see, the concrete conditions of materializing their promises in the field brought many setbacks to their triumphalist rhetoric.

It is certainly not a coincidence that to explore in detail one of the major plans that would change the Portuguese landscape in the 20th century we are forced to look once more at colonialism. Trigo de Morais, the engineer responsible for the 1935 Hydraulic Plan, also started his career in Mozambique. His trajectory, as well as the one of Ezequiel de Campos, suggests again how Africa had become an important proving ground for large-scale technological experiments later applied in the metropole. Before finishing his degree in civil engineering at the Technical Institute, he designed in 1919, together with the agricultural engineer Ruy Mayer, an irrigation project for the Buzi Company, one of the chartered companies to whom the Portuguese government had outsourced the colonization of Mozambique.[18] The Berlin Conference of 1885 that, as detailed above, officially launched the scramble for Africa, granted Portugal formal control over the territory of present day Mozambique. Nevertheless, Portuguese presence in the territory was always limited (it reached some 18,000 people in 1930), ceding large plots of land to chartered companies formed by international capitals, such as the Buzi company, with discretionary power inside concessionary areas. Most of the income of these companies derived from extracting taxes from African populations living under their domain and exporting conscripted labor to South African gold mines or Katanga copper mines. The economy of Mozambique was totally dependent on its neighbors, with railways transporting ores from South African and Southern Rhodesian mines to be shipped at the Mozambican ports of Lourenço Marques and Beira, and returning in the opposite direction carrying conscripted workers to work in the mines. Salazar's New State would demand more from the colony, urging it to provide raw materials and markets for metropolitan industries.[19]

---

[18] P. António da Silva Vieira, *Trigo de Morais Segundo Alguns Depoimentos*, (Lisboa: Agência Geral do Ultramar, 1967); A. Trigo de Morais, *Para uma Política Hidráulica e de Colonização em Moçambique*, (Porto: 1a. Exposição Colonial Portuguesa, 1934).

[19] José Negrão, *Cem Anos de Economia da Família Rural Africana*, (Maputo: Promédia, 2001); Malyn Newitt, *A History of Mozambique*, (London: Hurst, 1977); William Gervase Clarence-Smith, *O III Império Português, 1825–1975*, (Lisboa: Teorema, 1985); Patrick Harries, "Labor migration from Mozambique to South Africa: With special reference to the Delagoa Bay hinterland, 1862–1897," (Ph.D. dissertation, SOAS, 1983); Leroy Vail and L. White, *Capitalism and Colonialism in Mozambique: A Study of Quelimane District*, (London: Heinemann, 1980); José Capela, *O Imposto de Palhota e a Introdução do Modo de Produção Capitalista nas Colónias*, (Porto: Afrontamento, 1977).

Nervous and enthusiastic young army officers insistently warned against the danger of occupation by rival European powers of the country's African possessions due to the scarce Portuguese presence on the ground. These officers would be instrumental in supporting Salazar's dictatorship, which assumed as one of its main tasks the "nationalization of the empire."[20] The phrase meant that not only Portuguese national life would be geared toward the empire, but also that life in the colonies should have Portugal as its main referent. In 1930, the Colonial Act, issued when Salazar assumed temporarily the post of Minister of the Colonies, solemnly institutionalized the relations between Nation and Empire: "It is in the organic essence of the Portuguese Nation to undertake the historical function of possessing and colonizing overseas domains and to civilize indigenous populations."[21] One could find this same imperial rhetoric among the leaders of the previous Republican regime, but the push for Empire would be much intensified under Salazar's dictatorship. In fact, the building of an imperial bloc would become one of the defining features of Portuguese fascism.[22]

The increased production in the Buzi company lands due to Trigo de Morais' irrigation project earned the engineer in 1925 the position of technical director. That same year he put forward his first settlement project: the Company delivered 200 hectares to each new Portuguese settler willing to cultivate the land.[23] The contract signed between the Company and the settler also stipulated the first would buy the entire production of the latter paying 50% of its market value. By 1932, Morais was proud to announce that he had been able to establish 22 Portuguese families in the Buzi Company lands, showing how the empire could be nationalized and secured through white settlement. He was nevertheless vocal about the impossibility of scaling up the experiment based exclusively on private initiative. The state should take control of colonization efforts, "guiding, orientating and giving assistance to the first settlers," for, as he stated, "the colonization problem is but an agrarian problem, which can only be solved by Hydraulic Politics."[24]

Hydraulics was to be the solution to overpopulation occupying so many European engineers in the thirties. While Nazis in Germany set their *Östpolitik*, their imperial expansion to the East, in large measure as a conclusion of demographical considerations,[25] in Portugal the population distribution problem previously identified by Ezequiel de Campos was to be solved by irrigation. At the Colonial Exhibition of 1934 in Oporto, one of the main venues for the

---

[20] Valentim Alexandre, "Ideologia, economia e política: A questão colonial na implantação do Estado Novo," *Análise Social*, (1993), pp. 1117–1136.

[21] *Acto Colonial*, Decreto 18570, 8/7/1930.

[22] Significantly enough, the long span of the regime would only end in 1974 with the red carnations revolution led by young military officers refusing to keep fighting independence movements in Guinea Bissau, Angola, and Mozambique. With the empire in question, the fascist regime finally disintegrated.

[23] The next year, the Company sent his technical director, Trigo de Morais, to India to study the great irrigation works of British colonial engineers in Punjab. See Vieira, *Trigo de Morais*, p. 33.

[24] Morais, *Para uma Política Hidráulica*, pp. 28–29.

[25] See Gotz Aly and Susanne Heim, *Architects of Annihilation: Auschwitz and the Logic of Destruction*, (Princeton, NJ: Princeton University Press, 2003). On the continuity between white settlement projects between the Italian, Portuguese, and German fascist regimes, see Saraiva, *Fascist Pigs*.

deployment of the colonial ambitions of the New State in which that map of "Portugal is not a small country" was first shown, Trigo de Morais insisted: "the most efficient solution both in the Metropolis and in the Colonial Empire for the placing of metropolitan demographical surplus lies in public works of land reclamation by water and in its multipurpose uses." Each new dam in the arid Portuguese south would constitute the opportunity to divide large properties into small homesteads providing enough resources for the sustainability of individual families, which as we saw constituted one of the organic units of the regime, while the colonies would divert Portuguese emigrants from the American promises and attract them to irrigated land in Africa.

This was in a nutshell Trigo de Morais' program the moment he was nominated head of the National Commission for Agriculture Hydraulic Works in 1933. That very same year the Commission prepared a Plan of Studies and Constructions, which would become the mentioned Plan of Hydraulic works.[26] A new map of Portugal actualized Ezequiel de Campos' proposals presenting regional population densities and depicting twenty projected irrigation areas that corresponded to a total of 106,000 hectares (262,000 acres). The Commission planned to have irrigated more than 400,000 hectares (988,000 acres) by 1950, corresponding to 5% of the national territory, which, according to Morais calculations, would be able to absorb some 100,000 people, thus solving 10% of the Portuguese population problem, quantified in a surplus of one million people.

For such an endeavor Morais reorganized completely the Commission, involving in each project all its departments (agriculture, geological, topographical, civil, etc.), instead of nominating small brigades for each new project. To those criticizing the large sums of money spent per hectare of irrigated land, Morais countered that the amounts were smaller than those practiced in Spain or the South African Transvaal, reminding also that in Portugal his Commission was not only in charge of building the dams and the main irrigation canals, but also of hydroelectric production, the secondary canals network, the preparation of the land, the building of new rural roads, and the construction of houses for irrigation surveillance men controlling water flows and responsible for maintenance works.[27]

What looked like technical questions had indeed strong political implications, as highlighted by the inauguration of two major dams of the hydraulics plan in 1949. Significantly, one of the dams was named after the engineer Trigo de Morais while the other honored the dictator Salazar, completed with a bronze effigy of the ruler placed against the concrete. The inaugural speech by the minister of Public Works suggested how the irrigation dam materialized Salazar's accomplishments as uncontested leader of the country:

> "The dam is a work that establishes order and security; that regulates and foments the economy; that offers work, revenues and welfare; that uses the country's own resources to develop the many diverse fields of activity of its population... Would there

---

[26] A. Trigo de Morais, *Sempre o Problema da Água*, (Lisboa: Agência Geral das Colónias, 1945).
[27] Morais, *Sempre o Problema*, p. 38.

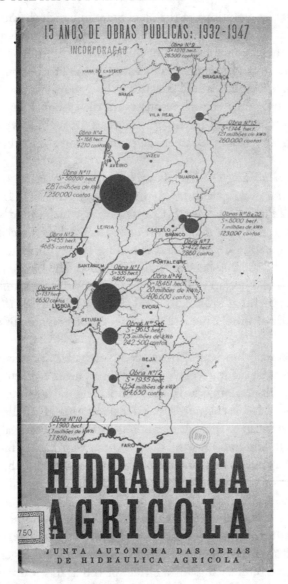

Figure 7.5: **Agriculture Hydraulic Works (1948)**.
The National Commission for Agriculture Hydraulic Works, directed by the engineer Trigo de Morais actualized Ezequiel de Campos' maps. Its 1935 plan promised to irrigate more than 400,000 hectares (988,000 acres) in fifteen years, corresponding to 5% of the national territory, which, according to Morais calculations, would be able to absorb some 100,000 people, thus solving 10% of the Portuguese population problem, quantified in a surplus of one million people. © Junta Autônoma das Obras de Hidraúlica Agrícola.

Figure 7.6: **Salazar Dam (1948).**
One of the first dams of the Irrigation Plan to be completed was named after the engineer Trigo de Morais while the other was named after Salazar, the dictator. In the inauguration, Salazar finished his speech with the call "Irrigate! Irrigate!"
© Public domain.

be any other material achievement, any other monument…which would better express the benefits derived from two decades of rule of the political regime inaugurated on the glorious day of May 28, 1926?"[28]

Salazar, present at the ceremony, answered with his own speech. The dictator recalled his early modest life as a small farmer, noting the importance of a regular water supply for those cultivating the land. But he emphasized that "water won't be only a factor of more abundant production… it will be as well the health of the people, the joy and freshness of the landscape." He finished his speech with the call "Irrigate! Irrigate!" The implication was clear: Green irrigated fields promoted the settlement of the south with small landowners, the nuclear families that constituted the basis of a national community abiding by the regime's trilogy "God, Fatherland, and Family."[29]

[28] *Diário de Notícias*, May 30, 1949.
[29] This trilogy was also the one used by the French Vichy regime, which shouldn't surprise us considering that its leader Marechal Pétain openly quoted Salazar's New State as its main source of inspiration.

Despite the rhetoric, by 1974—the end of Portugal's fascist regime—only 76,000 hectares had been irrigated.[30] Furthermore, irrigation failed to draw settler families into the large properties of the south. Instead, just the opposite occurred. Large landowners kept their estates undivided relying on underpaid seasonal wage labor and profited from State investments in an expensive hydraulic infrastructure that they never paid back. The principles of the hydraulic plan had been first exposed by Ezequiel de Campos, directly inspired by the American example of federal water control under the Bureau of Reclamation, and it suffered from similar failures.[31] In the American West, instead of supporting small farmers, massive irrigation works served to subsidize well-established agribusiness.[32] In Portugal, the growing infrastructural power of the New State first and foremost benefitted large landowners who constituted one of the supporting elites of the regime. The Hydraulic Plan's supposed beneficiaries had no political representation and censorship of the press guaranteed that criticism of public policies never reached the public sphere. Instead, the New State's propaganda machine publicized figures associated with the new dams as if everything happened according to the plan.

But the irrigation dams designed by the Commission engineers did bring major changes. One just has to look to the Portuguese food supply, namely the huge increase in rice production that earned that grain the status of a key staple in the national diet. The 20,000 tons of rice produced between 1925 and 1929 rose to 156,320 tons in 1954 and 1958, making the country self-sufficient in that grain. The higher demands of water of rice in relation to other crops implied a lower total area of irrigated land than initially projected, but more importantly, rice also demanded a seasonal work force that, rather than settling in low density areas and forming new virtuous family units, travelled each year from north to south and back again after the job was done.

The engineer Trigo de Morais's praised the virtues of "irrigation expelling the communist plague and strengthening the conservative basis of the Nation by reclaiming land and dividing property."[33] Instead, irrigation in the south of Portugal led to increased proletarianization of the rural workforce. The third way envisioned by fascist ideologues, with rural families of small landowners refusing communist promises of abolition of private property while benefiting from the division of large properties promoted by the state never occurred. The southern regions in which dams were built would in fact turn into one of the main scenarios of opposition to the regime from the 1950s onward with growing presence of the communist party among landless rural workers.

---

[30] Such numbers include as well those of the Alentejo Irrigation Plan launched in 1957. Fernando Oliveira Baptista, *A Política Agrária do Estado Novo*, (Porto: Afrontamento, 1993), p. 80.

[31] For the direct relations between Portuguese civil engineers serving the New State and the American Bureau of Reclamation see Tiago Saraiva, "Laboratories and landscapes: The fascist new state and the colonization of Portugal and Mozambique," *Journal of History of Science and Technology*, 3, pp. 1646–775.

[32] Worster, *Rivers of Empire*; Donald J. Pisani, *Water and American Government: The Reclamation Bureau, National Water Policy, and the West, 1902–1935*, (Berkeley: University of California Press, 2002).

[33] Cabral, *Materiais para a História*, pp. 102–106.

Trigo de Morais would have to travel once again to Africa to try to turn into reality his dreams of irrigation sustaining Portuguese virtuous families. In Southern Angola, the Cunene River basin represented a sort of last frontier for Portuguese colonizers. According to Morais, the region was waiting for a Christianization mission by Portuguese white settlers. Settlement villages on the Cunene banks were to be occupied "with the largest possible number of white settlers from the Metropolis…well rooted in the soil, owners of the land they work and carrying out the traditional virtues of the Portuguese farmer—stiffness, sobriety, working pride, an example for their African brothers."[34]

In 1946 the National Commission for Agriculture Hydraulic Works sent a mission to the south of Angola "to study the use of the Cunene waters in order to establish the likelihood of irrigating the territory, supply the energy needed for chemical fertilizers production, and to allow the fixation of European population in that very important economical and political zone."[35] The head of the Angola mission was Morais himself who after applying his Mozambique experience to the colonization of Portugal had finally been able to involve State resources in white settlement irrigation projects in Africa.

The Mission to the South of Angola presented an ambitious plan of 12 dams irrigating 42,500 hectares and reclaiming a total of 420,000 additional hectares of pastures for cattle, and producing 309 Gwh of energy. The Five Year Foment Plan (1951–1955), the major budgetary instrument of Salazar's government to allocate public investment after the afore mentioned Law of Economic Reconstruction of 1935, included funding for irrigation and drainage of the Cunene valley, preparation of terrain, installation of settlers, technical assistance, and, crucially, the building of the Matala dam, the first one of the twelve dams included in the plan. The Cunene works represented no less than a third of the total amount of funds of the Foment Plan for overseas territories.

As in so many other multipurpose dams, irrigation would progressively loose importance while electricity production assumed growing relevance. The settlements designed by Trigo de Morais, which were supposed to absorb some 3,000 Portuguese families, never housed more than 350.[36] The direct transfer of Portuguese agricultural practices into the high plains of Southern Angola, with the election of rice, tomato, and wheat as main crops of the new settlement, proved to be a complete disaster.[37] Although one of the model settlements of the irrigated area was named Vila de Folgares, homage to the Portuguese rural village where the engineer Trigo de Morais was born, the reproduction of a mythical rural Portuguese community in Angola was a failure. Both in the south of Portugal and in the south of Angola, the dreams of engineers of

---

[34] P. António da Silva Vieira, *Trigo de Morais Segundo Alguns Depoimentos*, (Lisboa: Agência Geral do Ultramar, 1967).

[35] Rui Sanches, *O Aproveitamento do Rio Cunene. Sua Importância Internacional e para o Sul de Angola*, (Lisboa: LNEC, 1999).

[36] For a detailed informed study of white settlement projects in Portugal and Mozambique see, Castelo, Cláudia, *Passagens para áfrica: O povoamento de Angola e Moçambique con naturais da metrópole, 1920–1974*, (Porto: Afrontamento, 2007).

[37] Mariano Feio, *As Causas do Fracasso da Colonização Agrícola de Angola*, (Lisboa: Instituto de Investigação Científica Tropical, 1998).

accomplishing their "social role" as makers of communities of virtuous nuclear families sustained by engineering infrastructures never materialized.

## 7.3    ARCH DAMS AND AMERICAN MODELED RESEARCH ENGINEERS

Trigo de Morais endeavors exemplify the combination of technical expertise and traditionalist social and cultural values promoted by Salazar's fascist New State, a corporatist regime based on alleged organic social units of producers and nuclear families. While population densities and distribution constituted in the 1930s main arguments for making irrigation dams cohesive elements of Portuguese life, energy shortages during World War II made hydroelectricity the core of engineers' promises of the following decades. This chapter details how the new challenges brought by the building and design of new types of dams led to the formation a new type of engineer in the decades after World War II until the end of the dictatorial regime in 1974. Far from Morais' settlements, the habitat of this engineer was the laboratory. He merged engineering with research and took American institutions as major source of inspiration.

In 1944, still under the difficult conditions of the War, a National Electrification Law was passed by initiative of the sub-secretary of Energy and Industry, Ferreira Dias Júnior, who had trained at the Technical Institute as mechanical and electrical engineer. The law envisioned large dams as the first energy source of the country. The year after, the state intervened directly in the creation of two large hydroelectric companies to develop the basins of the Cávado and Zêzere rivers, and in 1953 of a third one to develop the Douro Valley.[38] In 1946, construction started for the large dams of Castelo de Bode in the Zêzere River and Venda Nova in the Cávado River. The massive investments involved in these companies were justified with the creation of heavy industries such as electrochemistry and electrometallurgy, in a typical import substitution industrialization policy that fitted well with the nationalistic characteristics of the regime by relying on the country's own capacities for industrial development.[39]

Technical Institute engineers were of course at the center of the whole endeavor: Civil engineers applied their expertise in dam design and building; chemical engineers found a new market in chemical plants profiting from electricity availability such as Amoníaco Português (Estarreja) and União Fabril do Azoto (Alferrarede); mechanical engineers were in high demand by Mague and Sorefame, metalworking companies growing together with the new scale of the large dams construction sites; electrical engineers were key to put in place and manage a national

---

[38] *Direcção Geral dos Serviços Eléctricos, Concessão do Aproveitamento Hidroeléctrico da Energia das Águas dos Rios Cávado e Rabagão*, (Lisboa: Imprensa Nacional, 1964); Ilídio Mariz Simões, *Pioneiros da Electricidade em Portugal e Outros Estudos*, (Lisboa: EDP, 1997); Maria Fernanda Rollo and J. M. Brandão de Brito, "Ferreira Dias e a constituição da companhia nacional de electricidade," *Análise Social*, 31, (1996), pp. 343–354.

[39] Brandão de Brito, "Os engenheiros e o pensamento económico do estado novo," in *Dicionário Histórico de Economistas Portugueses*, Ed., José Luís Cardoso. (Lisboa: Temas e Debates, 2001), pp. 223–225; Nuno Madureira, "Visionários e dirigentes" Visionários e dirigentes: Os engenheiros portugueses na primeira metade do século XX," Comunicação apresentada ao *XX Encontro da Associação Portuguesa de História Económica e Social*, APHES, Faculdade de Economia do Porto, 24–25 de Novembro de 2000, pp. 16–17.

Figure 7.7: **Portuguese Exhibition in Lausanne (Switzerland), (1957).**
Posts, cables, and substations of the National Company of Electricity distributed power throughout the Portuguese landscape, regulating the development of the country according to engineers' plans.
© Biblioteca de Arte da Fundacao Calouste Gulbenkian.

electrical power network. If the civil engineer Trigo de Morais embodied with his agriculture irrigation projects much of the first phase of the New State regime, the mechanical and electrical engineer Ferreira Dias would be the exemplary figure of the second world war aftermath and its new emphasis on heavy industry.

Dias' vision of Portuguese rivers powering heavy industries all integrated through a national electrical grid had its most clear translation in the National Company of Electricity (CNE) founded in 1947 and headed by Ferreira Dias himself. The state-owned company's high voltage infrastructure connected the hydropower plants in the Cávado, Zêzere, and Douro river basins with the main consuming centers in the coast, namely the two main cities of Lisbon and Oporto, but also with the new industrial poles such as Estarreja and its chemical plants. The posts, cables, and substations of CNE distributed electricity throughout the Portuguese landscape, regulating the whole development of the country according to the engineer's plan.

From 1951–1972, 12 large dams were built, responsible for producing no less than 90% of the country's electricity by the end of the 1960s.[40] Considering the typical irregular Mediterranean regime of Portuguese rivers, with winter torrential flows giving place to reduced streams in the summer months, big structures and large artificial lakes were mandatory to guarantee a steady energy supply.[41] The use of traditional methods of dam design based on the mass of the structure to hold waters back—gravity dams—would have meant that most watercourses of the country would have remained unexplored. The enormous amount of cement demanded by such structures impeded the transformation of Portuguese rivers into a national resource, due to their excessive costs. Thinner and cheaper structures such as arch dams (in which water pushes into both foundations and abutments) were needed to materialize visions of Portuguese resources supplying energy for an industrial surge. A new kind of engineer would emerge together with these new structures—the research engineer.

In 1942 the Center for Studies of Civil Engineering (Centro de Estudos de Engenharia Civil) was founded and then renamed four years later as National Laboratory of Civil Engineering (Laboratório Nacional de Engenharia Civil—LNEC, from here onward National Laboratory).[42] The laboratory started to operate in the premises of the Technical Institute and was directed by Manuel Rocha, a young engineer who had recently returned from the Massachusetts Institute of Technology (MIT) where he became interested in researching materials' resistance.[43] Rocha's American experience was made possible by a scholarship from the Institute for High Culture (Instituto de Alta Cultura), created to support the training of the New State scientific elite.[44] In addition to offering young talented scientists the opportunity of attending renowned international universities, the Institute funded local centers of excellence in Portugal to welcome back the young scientific talents. The National Laboratory started as one of the Institute's centers. In a tiny laboratory six young and enthusiastic civil engineers were starting a research program that would become the most successful Portuguese experience with large-scale research.

The National Laboratory would centralize all experimental activity connected with civil engineering problems, responsible not only for the study of structures and materials, but also for defining standards and developing construction methods.[45] But Manuel Rocha insisted that

---

[40] For a general overview of dam building in Portugal, see Rui Candeias Jacinto, "As barragens em Portugal: De finais de oitocentos ao limiar do século XXI," in *Momentos de Inovação e Engenharia*, Eds., Manuel Heitor, J. M. Brandão de Brito, and M. Fernanda Rollo, vol. II, (Lisboa: Dom Quixote, 2004), pp. 800–819.

[41] Carvalho Xerez, "Aproveitamento hidroeléctrico dos nossos rios. Estudos realizados pela direcção geral dos serviços hidráulicos," Paper presented at the *Ordem dos Engenheiros*, Lisbon, May 1945.

[42] *O Laboratório de Engenharia Civil. Organização e Funcionamento*, (Lisbon: Ministério de Obras Públicas, 1949); Manuel Rocha and J. Laginha Serafim, "Model tests of Santa Luzia dam," *3rd Congress of Large Dams*, Stockholm, 1948.

[43] For a biographical account of Manuel Rocha see, Armando Gibert, *Manuel Rocha, O Pensamento e a Obra*, (Lisboa: Imprensa Nacional Casa da Moeda, 1986).

[44] On the activities of the Institute of High Culture see Fernanda Rollo, *Ciência, Cultura e Língua em Portugal no Século XX. Da Junta de Educação Nacional ao Instituto Camões*, (Lisboa: Imprensa Nacional-Casa da Moeda, 2012); Júlia Gaspar, Maria do Mar Gago, and Ana Isabel Simões, "Scientific life under the Portuguese dictatorial regime, 1929–1954: the communities of geneticists and physicists."

[45] Manuel Rocha, *The Laboratório Nacional de Engenharia Civil. Organization Management and Activity*, (Lisbon: LNEC, 1955), pp. 1–2.

Figure 7.8: **Santa Luzia Arch Dam**.
The use of traditional methods of dam design based on the mass of the structure to hold waters back—gravity dams—would have meant that most watercourses of the country would have remained unexplored. Only thinner structures such as arch dams (in which water pushes into both foundations and abutments) were able to materialize visions of Portuguese resources supplying energy for an industrial surge. A new kind of engineer would emerge together with these new structures—the research engineer.
© Biblioteca de Arte da Fundacao Calouste Gulbenkian.

the importance of the laboratory for the nation was first and foremost its research on arch dams. Engineers recruited among the best students from the Technical Institute, tested different qualities of concrete; observed structural deformations, as well as stresses and temperatures during and after dam construction; examined rock foundations; or carried model studies of dams.[46] In fact, model studies of arch dams would become the most distinguishing feature of the research undertaken at the National Laboratory, proving to be an effective tool to make it participate in the national electrification effort as well as to gain international recognition.

---

[46] Rocha, *The Laboratório*, pp. 17–21.

Figure 7.9: **Manuel Rocha**.

Manuel Rocha was trained at the Technical Institute in Lisbon and at the Massachusetts Institute of Technology (MIT) where he became interested in materials' resistance. Rocha's American experience was made possible by a scholarship from the Institute for High Culture (Instituto de Alta Cultura), which supported the training the New State scientific elite. The National Laboratory of Civil Engineering started as one of the Institute's centers. In a tiny laboratory, six young and enthusiastic civil engineers started a research program that would become the most successful Portuguese experience with large-scale research.
© Public domain.

In 1944, no more than two years after its creation, the research engineers started to work on a first model of the Santa Luzia dam, the first arch dam built in the country.[47] Starting from the United States Bureau of Reclamation experience with models of Hoover Dam, a research program was launched by Manuel Rocha to systematically use small physical models in arch dam design.[48] The aim was to overcome the high manpower and time requirements of American analytical methods of dam design by developing modeling techniques that would also improve accuracy in establishing the stress state of the dam.[49] Hundreds of engineers had been mobilized to the offices of the Bureau of Reclamation in Denver, Colorado, to perform the calculations

[47] Rocha and Serafim, "Model tests."
[48] See for example, Manuel Rocha et al., "Model tests, analytical computation and observation of an arch dam," *Proc. of the American Society of Civil Engineers*, 81, (1955), Separate no. 696.
[49] Manuel Rocha and J. Laginha Serafim, "Analysis of concrete dams by model tests," *5th Congress of Large Dams*, Paris, 1955.

involved in the design of the Hoover dam. Models were used only as an auxiliary test tool to probe the designs produced by analytical methods. The masses of engineers arriving in Denver from the American engineering schools were impossible to reproduce at the Portuguese scale. The example was certainly coming from the United States, but this had to be reworked to be viable in Portuguese conditions. The National Laboratory could not grow around the design methods developed in the U.S. and would focus its efforts on original research on physical models instead.

The limits imposed to engineering research in Portugal determined by the size of the engineering profession led thus to the emergence of laboratory methods of dam modeling: different model materials were tested, several methods of reproducing the load on the structure were tried, and new instruments for the measurement of deflections were developed. In testing models, the accurate measurement of small deformations was a crucial issue, which explains why Manuel Rocha insisted on the importance of counting with an active Instrumentation Section at the National Laboratory. Electrical extensometers were developed to replace traditional vibrating chords, which demanded improved facilities to control experiments' conditions. Measurements made at the National Laboratory were then compared with field data from the real dams which were equipped with a set of instruments following an observation plan also designed by the laboratory in tight collaboration with the building companies.[50]

Manuel Rocha and fellow engineers from the Department of Structures of the National Laboratory, claimed in several papers presented to the International Congresses of Large Dams and published in the Proceedings of the American Society of Civil Engineers, to have demonstrated the superiority of models when compared to analytic tools in offering a better account of the complexities inherent to arch dam structures.[51] According to their experiments, the use of models was in fact mandatory for designing safe thin structures, especially in sites of irregular profile, in dams with singularities (such as spillway openings), and in the case of heterogeneous foundations. Or, more bluntly, models were mandatory for any important structure. All Portuguese large arch dams' designs started to be tested at the National Laboratory premises before laying the first stone. In contrast to the U.S. Bureau of Reclamation practice, which only used models to confirm the designs coming out of analytical methods' computations, the Lisbon laboratory put all its efforts in developing cheap and flexible model techniques that earned its international reputation. In the beginning of the 1960s the very same U.S. Bureau of Reclamation was hiring the Portuguese laboratory to make the model studies of the complex Morrow Point Dam structure in the Gunnison River in the Colorado basin.

In 1955, 11 years after the beginning of the first model of the Santa Luzia dam, Manuel Rocha could already argue that the experimental technique developed at the National Laboratory, through its direct savings in structural costs, had already compensated for the entire

---

[50] Manuel Rocha, J. Laginha Serafim, A. Ferreira da Silveira, and O. V. Rodrigues, "The observation of the behavior of the Portuguese concrete dams," *5th Congress of Large Dams*, Paris, 1955.

[51] Rocha et al., "Model tests, analytical."

Figure 7.10: **Testing Dam Models at the National Laboratory of Civil Engineering (LNEC) (1962).**

Starting from the United States Bureau of Reclamation experience with models of Hoover Dam, a research program was launched by Manuel Rocha at LNEC to systematically use small physical models in arch dam design. PLNEC research engineers overcame the high manpower and time requirements of American analytical methods of dam design by developing modeling techniques.

© Biblioteca de Arte da Fundacao Calouste Gulbenkian.

investment made by the State in its facilities.[52] Such cost-benefit analysis was not an innocent claim for we are dealing with a laboratory that started in the 1940s with 6 young engineers and improvised research facilities, and that grew in 1954 into a national laboratory employing 490 people, from which 90 were research engineers. The workforce numbers demanded a new organization of research as suggested by Rocha's description of his own institution: "the laboratory has an industrial type organization which enables it to determine the real cost of each service after its conclusion. Each member of the staff has a card in which he records daily the time spent in the different jobs, discriminating even the time spent in studies, in consultations, in receiving visitors…"[53] While the National Laboratory promised to optimize construction techniques across the country, research in its interior was also to be optimized.

Research activities, according to Rocha, were cautiously planned to avoid "doing research for the sake of research. The criterion for the choice of a problem is the service it renders to the country."[54] The new facilities inaugurated in 1952 reveal Rocha's idea of what a national research institution ought to be. It is evident that Rocha worked closely with the architect Pardal Monteiro, who had also designed the Technical Institute facilities described above. The laboratory monumental character was in no contradiction with the simplicity of the concrete façade, an obvious homage to the work being carried behind the gray walls. A quick look is enough to realize the modular nature of the building, with no differentiation of the several sections of the laboratory.[55] Each modulus of 7 × 3.5 m was limited by glass walls easily removable in case of necessity. The laboratory was perceived as a research machine easily adoptable to new objects of inquiry requiring different space distributions for teams and instruments. Nevertheless, the clear hierarchy of the institution was not forgotten in this modular structure. Each section was distributed in a single zone of the building in a cascade occupying its three floors, in order for the engineer director of the section to supervise the subordinates' work.

The National Laboratory was located in the northern border of the city. The urban space between the Technical Institute and the National Laboratory, most of it developed between the 1930s and the 1950s was a true showcase of the building capacity of the New State and of its different phases. Closer to the Technical Institute, the Londres and Areeiro squares as well as the Afonso Henriques Parkway combined the new building scale introduced by concrete in the 1930s with historicist references to Portuguese imperial past, a peculiar combination forming what historians of architecture know as the "soft Portuguese style" of the New State, similar to the styles characterizing Mussolini, Hitler, and Franco regimes. Further north, in an area limited by the National Laboratory and the Julio de Matos Hospital, the Alvalade neighborhood, started in the aftermath of World War II, materialized the social policies of the New State. Cheap

[52] Rocha, *The Laboratório*, p. 2.
[53] Rocha, *The Laboratório*, p. 3.
[54] Rocha, *The Laboratório*, p. 2.
[55] We follow the unpublished architecture project for the laboratory building, Pardal Monteiro, *Laboratório Nacional de Engenharia Civil. Programa e Memória Descritiva. Ante-Projecto*, (Lisbon: Ministerio de Obras Públicas, 1948).

Figure 7.11: **The building of the National Laboratory of Civil Engineering.**
The LNEC facilities inaugurated in 1952 were designed according to Manuel Rocha's demands for a modular flexible building that could also keep hierarchy. The laboratory was a research machine easily adoptable to new objects of inquiry requiring different space distributions for teams and instruments. Each section was distributed in a single zone of the building in a cascade occupying its three floors, in order for the engineer director of the section to supervise subordinates' work.
© Photo by António Passaporte. Arquivo Fotografico da Camara Municipal de Lisboa.

housing financed by the different corporatist structures was built using standardized construction methods and prefabricated materials, some of them tested at the National Laboratory.

Well-planed neighborhoods in the capital city, large arch dams, a national electrical power network, heavy industry (nitrates, steel, shipyards, paper pulp, cement), a world-class engineering research laboratory; all affirmed the main role played by engineers in the making of the New State and their contribution to the golden years of Portuguese economic development starting in the end of the 1950s until the oil crisis of 1973. In 1963 Western Europe welcomed its most recent member to the club of industrialized countries: Portuguese industry contribution to the national GDP finally surpassed agriculture. Also, for the first time in modern history, the gap with northern Europe seemed to be closing up, with Portuguese economy growing faster than the European center during the 1960s. In 1960 the country joined the European Free Trade Association (EFTA), a major step in the internationalization of its economy.

The repeated celebration of engineers' accomplishments by the politically controlled mass media concealed a more troubled balance. The combination of large arch dams supplying energy for heavy industry supported an import substitute policy that had internal markets as its obvious limit. The highly protected large industrial companies employing hundreds of engineers had been the main contributors for the rapid growth of the country's industrial product since the 1950s. But the more dynamic sector profiting from the opening of the economy through EFTA and that would increasingly contribute with higher shares to the growth of the GDP through exports was the traditional textile industry.

It is important to retain that no engineers were employed by textile companies, which relying on the low wages of the Portuguese workforce would be responsible for the success story of the integration of Portuguese economy first in EFTA and later in the EEC European Economic Community. By 1970, the secretary of industry Rogério Martins, recognizing the limits of the import substitution policy that had guided engineers' imagination since the beginning of the century, demanded a "quick change of lane for a deep industrialization."

Engineers certainly changed the country but the benefits of their actions were socially limited. Municipal powers were vocal about the lack of benefits of large dams for local populations: while high voltage cables transported the energy that developed the industrial poles of Ferreira Dias' charts, the villages next to the dams lacked electrical lighting.[56] It is certainly revealing that Manuel Rocha, the head of the National Laboratory, would participate in the first government coming out of the April Revolution of 1974 that finally overthrew the dictatorship as minister of social equipment and the environment, the new name for the ministry of public works. In the revolution aftermath, brigades of volunteers would not only participate in literacy campaigns, as would bring as well to large parts of the territory electricity and running water. Yes, Portugal was growing faster than the rest of Europe in the 1960s, but it was still the Western European country dedicating a lower share of its GDP to social welfare.[57] The neighborhoods between the Technical Institute and the National Laboratory were showcases of the engineering profession, but in the city outskirts slums with no infrastructure provided the conditions for the return of cholera to Lisbon in the revolutionary year of 1974.[58] New engineers in the 1970s and 1980s and new institutions would emerge proposing a new relation between Portuguese democracy and technology.

[56] *Um município pobre contra o poder económico: a recusa da hidro-eléctrica do Cávado em pagar os impostos devidos à câmara de Montalegre ou um direito incontestável a que se pretende opôr-se uma desculpa incompreensível,* (Montalegre: Câmara Municipal de Montalegre, 1961).

[57] Luciano Amaral, *Economia Portuguesa, as Últimas Décadas,* (Lisboa: Fundação Francisco Manuel dos Santos, 2010).

[58] Tiago Saraiva, Luísa Schmidt, and João Pato, "Lisbon water regimes: Politics, environment, technology and capital, 1850–2010." *Flux,* 3, (2014), pp. 60–79.

CHAPTER 8

# Conclusion

## 8.1  REVOLUTIONARY YEARS BETWEEN TECHNOCRACY AND DEMOCRACY

In the 1960s, the *Estado Novo* would face major challenges, namely from national independence movements in Africa (Angola, 1961, Guinea, 1963, Mozambique, 1964) fighting against the last European empire in the continent. The colonial wars would lead to the final demise of the dictatorial regime in the April revolution of 1974, when low-rank officials of the Portuguese army took arms in name of democracy, decolonization, and development. University students also organized numerous demonstrations against the regime in the 1960s, joining the international wave of students' contestation epitomized by the Parisian May 68 and the mobilization against the Vietnam War in the United States. At the Technical Institute, students' opposition to dictatorship and the war in Africa, was combined with prosaic demands for more scholarships, students' housing, the ending of hierarchical relationship between students and professors, and more autonomy of universities from the state. Acknowledging the limited reach of students' demonstrations in opposing the regime, many Technical Institute alumni assert nevertheless their importance in the formation of a political commitment to democracy among engineering students.[1]

The Technical Institute campus and its students' union in the 1960s became indeed a privileged place to experiment with alternative sociabilities to the hierarchical and policed world of the Portuguese capital under the dictatorship. Student meetings, café afternoons, theater performances, literary conferences, music concerts, film sessions, were all occasions to embrace values and fashions arriving mostly from other European countries, north of the Pyrenees. Touristic tours organized by the students' union became an invaluable way of breaking the international isolationism imposed by a political regime that resisted democracy after the defeat of fascism in World War II. If we keep in mind the cultural ambitions of the engineering school when it changed its name from Industrial Institute to Technical Institute in 1911, maybe we won't be surprised by the profusion of those alternative cultural initiatives in the 1960s and the 1970s. But the fact is while in the 1910s engineering, through the German notion of *Technik*, was being elevated to the condition of constitutive element of culture, in the 1960s cultural wars what was at stake was how to be an engineer who also was cultivated in art and politics. This was not anymore a question of being cultivated and politically active through the practice of engineering,

---

[1] Luísa Tiago de Oliveira and Marta Silva, "O ativismo estudantil no IST (1945–1980)," in Jorge Freitas Branco, Ed., *Visões do Técnico no Centenário, 1911–2001*, (Lisbon: ISCTE, 2013), pp. 307–397.

but of being an engineer who also had an interest in culture, politics and society. The cultural questioning of the 1960s didn't seem to include questioning what engineering was.

Some isolated events seem to indicate alternative possibilities. Excursions put students in contact with the social realities of a very unequal country that had little correspondence with the well-ordered technocratic foment plans drafted by their elders. The catastrophic floods of the fall of 1967, in which some 500 people lost their lives in areas close to the capital city, were perceived as a call to action by many engineering students. Indeed, it was the Technical Institute students' union that organized the disaster relief actions put in place by the whole community of Lisbon university students. After this traumatic event, students' demands routinely included mentioning of making engineering more responsive to populations actual needs. When the democratic revolution finally came in 1974, engineering students would indeed participate in "student civic service" activities, bringing sewage, electricity, and literacy to underserved communities across the country.

These initiatives deserve further studying to understand their actual reach, but it seems safe to affirm that while many engineers were active political actors of the turbulent years of the revolution aftermath, they didn't tend to express their democratic political visions through new forms of engineering, as engineers had done in the liberal and fascist periods. Training at the Technical Institute gave them the cultural capital needed to join the new political elite, but this had no apparent relation to their engineering practice, still associated in large measure to industries that prospered in the dictatorship years.

Maria de Lurdes Pintassilgo is an eloquent case in point: trained as chemical engineer at the Technical Institute she would be employed in the 1960s by the large chemical conglomerate of CUF while starting a parallel social activism connected to progressive catholic circles and reformers of the regime. After the revolution, she would continue to be a prominent political voice for the building of an extended welfare state, first as secretary and minister of three provisionary governments (1974/1975) and later as prime minister (1979/1980) of the fifth constitutional government.[2] Saint Simonians expressed their social utopia through railways and industries; fascists could imagine their alternative modernity through affordable housing and irrigation dams; and Pintassilgo's social welfare initiatives had no apparent relation to chemical engineering.

Another major Portuguese political figure, António Guterres, confirms the tendency: he graduated as electrical engineer from the Technical Institute in 1971, only to become in 1974 a full-time politician as member of the Socialist Party, later prime minister from 1995–2002, and currently no less than the secretary-general of the United Nations. Guterres certainly recognizes the relevance of the training he had at the Technical Institute, but his stellar political career didn't owe much to his engineering practice: always an exemplary student, he recalls how beyond excelling in mathematics or physics, he learned "an interest for other things, such as society, other

---

[2] Not surprisingly, the aftermath of the revolution of 1974 were turbulent political times with a quick succession of governments: six provisional governments between 1974 and 1976; nine constitutional governments from 1976–1985.

people, sports, cultural activities… [At the Technical Institute] it was the first time I was told there were things in life other than studying."[3]

Engineers who did owe their political career after 1974 to their professional practice as engineers suggest instead uncomfortable continuities between the fascist and democratic regimes on the perceived role of engineering in Portuguese society. The Ministry of Industry of the democratic regime, namely, was largely dominated until the 1980s by engineers who made their careers in the large industrial conglomerates that had grown under the protection of the dictatorial state.[4] Alfredo Nobre da Costa, for example, another Technical Institute alumnus, was a prominent member of the Portuguese engineering elite having worked since the 1950s for the Champallimaud group (cement and steel works), EFACEC (electrical equipment), and SACOR (oil refinery). After a first experience after the democratic revolution as secretary of heavy industry of the second provisional government in 1975, he would become in 1977 minister of industry and technology of the first constitutional government, being promoted in 1978 to prime minister of a short-lived government from August to November 1979. A self-declared apolitical engineer, Nobre da Costa was presented as a transition solution to keep government going until alleged actual politicians could form a proper government.

The point here is not so much to dispute the democratic credentials of the ministerial elite of the first years of the democratic regime, as to call attention to how engineering was being presented as an apolitical activity that could transit without impediments from one political regime to the other. The flaws of such technocratic pose can be appreciated by briefly referring the dispute over the building of a nuclear power plant in Portugal.[5] This was a project developed under the dictatorship that promised a major domestic source of energy while making use of the abundant uranium ores of the country.[6] To produce the technical know-how necessary for the development of a nuclear sector, a new national laboratory—the Laboratory of Physics and Nuclear Energy—had been created in 1959, following the previous example of the National Laboratory of Civil Engineering. The nuclear plant was indeed thought of as device to multiply the technical elite that would lead the country to a new development stage, "to create…[and] educate scientists, to develop the first sector of high technological density in Portugal."[7] Under the major energetic challenge created by the oil crisis of 1973, engineers at the Ministry of Industry had no doubts in keeping pushing for nuclear energy in the aftermath of the democratic

---

[3] Quoted in Oliveira, p. 330.

[4] A 1972 survey of the engineering profession revealed that 54% of Portuguese engineers worked in private companies, and only 24% (the second most significant group) worked for the State. Significantly enough, 72% of Portuguese engineers worked in organizations of more than 200 employees, with 28% of them (the largest group) in organizations employing more than 2,000 people. Rodrigues, *Os Engenheiros*, pp. 128–129.

[5] Tiago Santos Pereira, Paulo F. C. Fonseca, and António Carvalho, "Carnation atoms? A history of nuclear energy in Portugal." *Minerva*, (2018), pp. 1–24.

[6] Maria Júlia Neto Gaspar, "Percursos da física e da energia nucleares na capital portuguesa: Ciência, poder e política, 1947–1973." Ph.D. dissertation, University of Lisbon, 2014; Tiago Saraiva, Ana Delicado, and Cristiana Bastos, "Configurações da investigação científica em Portugal: três estudos de caso." *Itinerários. A Investigação nos 25 Anos do ICS*, Ed., Manuel Villaverde Cabral et al., (Lisboa: Imprensa de Ciências Sociais: 2008), pp. 429–450.

[7] João Caraça, quoted in Santos Pereira, p. 511.

revolution. They never thought the issue could be politicized, or better said, they never realized their position as engineers was also political and thus open for debate in a democratizing society.

In March 1976, only two months after the Minister of industry Walter Rosa—an electrical engineer who had participated in the dam building effort of the previous regime—announced the beginning of the construction of the nuclear power plant near the coastal village of Ferrel, a popular mobilization started.[8] Following international patterns of environmentalist movements against nuclear energy in other democracies, dissident voices made themselves heard through demonstrations, public meetings, the press or music festivals. Engineers from the ministry tried to win demonstrators' minds by informing the lay public of the alleged sound scientific principles on which everything was planned. But opposing powerful arguments were put forward concerning possible impacts for fishermen communities of water temperature rise, the economic soundness of the project, or the seismic risk of the area. This last issue was crucial enough to put the project on hold: Portugal still has no nuclear power plant.

Importantly for this book, the counter arguments owed much of their political track to the direct involvement of Technical Institute young engineers such as José Delgado Domingues, who maintained a close dialogue with concerned populations, and José Matos Ferreira and Tito Mendonça who headed a "Comission for Promoting the National Debate on the Nuclear Option."[9] The active participation of engineers in this first Portuguese environmentalist movement, exemplified a possible alternative relation between engineering and society. In a new political regime that promised to democratize, develop, and decolonize, many engineers still imagined their role limited to the second "d". The nuclear story confirmed nevertheless how important engineers could be in building an actively democratic society.[10]

## 8.2   NEW DISCIPLINES AND INSTITUTIONS

Such democratic commitment demanded a deep transformation of engineering education. It is good to keep in mind that by 1960s, only 1.2% of the economically active Portuguese population had completed a university degree, and engineering was still a rather elitist course only taught at the Technical Institute in Lisbon and the Engineering Faculty in Porto. In 1972, there were some 11,200 engineers in the country and the two schools graduated every year some 300 students, a number that had been kept steady since the late 1940s.[11] Building on the students' contestation of the 1960s, the so-called "liberal wing" of the National Union Party, the single party of the

[8] Stefania Barca and Ana Delicado, "Anti-nuclear mobilisation and environmentalism in Europe: A view from Portugal, 1976–1986," *Environment and History*, 22, 4, (2016), pp. 497–520.

[9] Ana Delicado, "Scientists, environmentalists and the nuclear debate: Individual activism and collective action," in *Associations and Other Groups in Science: An Historical and Contemporary Perspective*, Ed., Ana Delicado, (Newcastle: Cambridge Scholars Publishing, 2013), p. 200.

[10] The role of engineers in the third "d"—decolonization—is a major subject totally unexplored by the historiography. Participation of Portuguese engineers in the independence projects of Angola, Mozambique, and Cape Vert are important topics waiting for further research.

[11] Rodrigues, *Os Engenheiros*, p. 136.

Figure 8.1: **The Technical Institute engineer Delgado Domingos addressing the local population and protesters of the project of a Nuclear Plant in Ferrel (Peniche, 1976).**
The active participation of engineers in this first Portuguese environmentalist movement, exemplified a possible alternative relation between engineering and society. In a new political regime that promised to democratize, develop and decolonize, many engineers still imagined their role limited to the second "d". The nuclear story confirmed nevertheless how important engineers could be in building an actively democratic society.
© Photo by José António Miranda. Ephemera, Biblioteca e Arquivo de Jose Pacheco Pereira.

dictatorial regime, called for a debate on the national system of higher education, which opened the doors for its massification.

José Veiga Simão, a physicist trained at Cambridge University and Minister of National Education from 1970–1974, the year of the democratic revolution, promoted a major reform of the whole system. The so-called Veiga Simão reform of 1970, inspired by the British policies of the 1960s that created 23 new universities in the country,[12] was based on the developmentist belief that higher education constitutes the driving force of economic and social transformations, with universities producing the skilled workforce feeding economic growth based on science and

---

[12] In 1963, the British government commissioned a report to the economist Lord Lionel Robbins, professor at the London School of Economics. As a result, between 1961 and 1969, Britain created 23 new universities and the number of students increased substantially reaching the threshold of massification in 1988.

technology. The reform did lead to a significant increase of the Portuguese university population: from 24,000 students in 1960–1961 to 44,000 in 1970–1971, and 61,000 in 1974–1975.

In 1973, Veiga Simão's ministry established in addition to polytechnic institutes (offering shorter three year courses), four new universities: Évora University Institute (converted in 1979 into a university); University of Minho; University of Aveiro; Nova University (New University) of Lisbon. With the important exception of the latter, the new institutions were located outside the cities of Lisbon, Porto and Coimbra and should thus contribute to anchor new territorial nodes of economic and demographic development.

The naming of the engineering school of Nova as Faculty of Sciences and Technology (FCT) reflects the growing importance of notions imported from the United States and England and the forgetting of the German model of *Technik* underneath the naming of the Technical Institute. Evidence-free linear models of innovation produced namely by American business schools now promised economic growth through technologies based on research in the fundamental sciences.[13] While the word "technology" had not been part of the vocabulary used to describe engineering education at the Technical institute, FCT now brought it to the forefront following the example of English language which only in the early 1960s started to commonly use the term, mostly as a synonym of applied science.[14] Probably more important than the naming, to establish itself as a new kind of engineering school, FCT invested in new engineering courses that did not exist at the Technical Institute nor at the Faculty of Engineering in Porto, as well as in new pedagogical methods that brought teachers closer to students. Also, FCT settled in the southern bank of the river Tagus (facing Lisbon), in one of the country's most industrialized areas (auto makers, shipyards, chemical industry, etc.), in a large campus very much inspired by American and British models. Its horizontal buildings and shared spaces were designed to promote the interaction between students and the surrounding community, a clear contrast to the monumental and imposing architecture of the Technical Institute.

FCT offered namely three new courses: Computer Sciences, Environmental Engineering (in collaboration with the University of Aveiro), and Industrial and Production Engineering. These courses, which eventually became FCT watermark, changed the panorama of engineering training in Portugal, allowing Portuguese engineering students not only to catch up with international trends but to have a different social presence in the country as well. Despite criticisms that doubted the long-term interest of these new engineering areas that some characterized as a fashion phenomenon as ephemeral as the mini-skirt, the course in Computer Sciences did take off in 1974, some ten years after the first courses in the United States and in Europe; in 1977, the first students enrolled in the course of Environmental Engineering at FCT based on a wider approach to the more traditional training in sanitary engineering, no more than fifteen years

---

[13] For a critique of this model from a history of technology point of view see, David Edgerton, "'The linear model' did not exist: Reflections on the history and historiography of science and research in industry in the 20th century," in *The Science-Industry Nexus: History, Policy, Implications*, Ed., Karl Grandin, Nina Wormbs, and Sven Widmalm, pp. 31–57, (Sagamore Beach, MA, Science History Publications, 2004).

[14] Schatzberg, *Technology*, pp. 212–213.

Figure 8.2: **Campus FCT NOVA (Monte de Caparica).**
The naming of the engineering school of NOVA as School of Sciences and Technology (FCT) reflects the growing importance of notions imported from the United States and England. FCT settled in the southern bank of the river Tagus (facing Lisbon), in one of the country's most industrialized areas (auto makers, shipyards, chemical industry,...), in a large campus very much inspired by American and British models. Its horizontal buildings and shared spaces were designed to promote the interaction between students and the surrounding community, a clear contrast to the monumental and imposing architecture of the Technical Institute.
© Public domain.

after the first course in the United States; in that same year the Bachelor degree in Industrial and Production Engineering, a comprehensive and interdisciplinary engineering degree, also took off, less than a decade after the first degrees in the United States, and answering directly to industry's needs.

Soon the differences between "old" and "new" universities, as far as courses were concerned, faded out and today all engineering schools in Portugal teach an array of courses that include the most traditional ones, such as civil or electrical engineering, those born in the 1970s, e.g., Computer Sciences and Environmental Engineering, and the more recent ones as, for instance, Biomedical Engineering or Nanotechnology Engineering.

## 8.3   SUMMARIZING THE ARGUMENT

It is easy to identify many flaws in our narrative. Too many institutions are missing, namely those outside of Lisbon such as the Oporto Polytechnic Academy or the Engineering Academy of Rio de Janeiro. We didn't explore the important tensions between engineers and other formally trained technical experts of lower social rank. Relevant episodes of Portuguese history with an obvious technological dimension such as the colonial wars of the second half of the 20th century were not even mentioned. Maybe more reproachable, this is a history of engineering Portugal that has largely left out women, workers, and colonized people. Yes, engineers were an elite of males for the periods under study, but the history of technology has already suggested many ways of incorporating other relevant social actors in engineering studies beyond those trained at engineering schools. We strongly believe that exploring new geographies, new historical contexts, and new historical actors constitute major challenges for the field of history of technology in the country, and we would like to encourage scholars to fill up some of those embarrassing gaps in the historiography. Acknowledging the limitations of the existent literature and our own shortcomings, we believe nevertheless that the field has produced enough scholarship to try a first synthesis of the history of engineering in Portugal as the one attempted here. The following closing paragraphs summarize it, emphasizing its importance for any proper understanding of the history of the country.

Seminal textbooks such as Serrão Pimentel's *Portuguese Method for drawing Fortifications* (1680) and Azevedo Fortes *The Portuguese Engineer* (17) defined what it meant to be a Portuguese engineer, in contrast to being just an engineer ready to be hired by a wealthy patron independently of national allegiance. These first Portuguese engineers trained in mathematics and in Italian, Dutch, and French military methods at the Class of Fortification became responsible for the design of an infrastructure of fortresses in the Portuguese territory and across its empire that guaranteed the viability of an independent country and a new dynasty. As they became involved in the project of glorification of the monarchy, engineers added to their military identity the role of courtier, appreciated not only by their capacity to build fortresses and direct sieges but also by their ability to design elegant urban scenarios such as fountained squares or scenic thoroughfares, or, even more unexpectedly, to discuss with eloquence in enlightened intellectual and scientific circles that established the historical legitimacy of the Portuguese monarchy. Engineers' designs became identified with the image of the monarchy itself as powerfully exemplified by the austere aesthetics of the rebuilt capital city after the 1755 earthquake.

The discontinuity of Portuguese state structures caused by the Napoleonic invasions and the civil war opposing liberals and absolutists meant a profound redefinition of what it meant to be an engineer. As with all other regime changes that followed, to imagine a new political configuration for the country implied imagining a new engineer. This was not made from scratch, and the previous system of making engineers that had worked through the entire 18th century was incorporated in the new Polytechnic School and the Army School, the institutions founded in 1837 by the liberal regime. But to understand the new social role expected from Portuguese engineers trained at both schools it is fundamental to refer the French example of the École Polytechnique and the engineering schools of application on which those two new schools were modeled. Building on the Saint Simonian technological utopia developed by their French counterparts in which communications, machinery and financial credit forged a new society, the polytechnic engineers would form the backbone of the Portuguese liberal state until the late 19th century. Engineers became themselves a powerful political elite, probably more so than in France, putting forward an agenda of "material improvements" that dominated the country's political agenda. Engineers' own beliefs, perceptions, and values were turned into a normative discourse making engineers into organic intellectuals, builders of a cultural hegemony and promoters of a technological-driven epistemology.

This golden era of Portuguese engineering would come to an abrupt end with the global financial crisis of the 1890s. Portugal defaulted on its debt and was not able to finance itself anymore in international markets halting the infrastructural expansion that was at the core of engineers' identity and unique position in Portuguese society. Portuguese engineers finally gave away their military identity, preferring instead to be identified as organizers of industry who would transform the nation through mines, electrification, and chemical plants. The Lisbon Industrial Institute would be reformed following the German model of the *Technische Hochschulen* and be renamed in 1911 Superior Technical Institute, translating into Portuguese the cultural concept of *Technik,* which offered engineers the same cultural status of other learned elites of the country. The institutional reform coincided with the founding of the new Republican political regime with engineers producing in the following years myriad of plans aimed at profoundly transforming Portuguese society and replacing, according to António Sérgio famous formulation, "transportation" by "fixation."

Most of those plans were implemented only after the coup of 1926 that gave place to an authoritarian regime ruling the country until 1974. Historians have insisted in the importance of the alleged victory of ruralists over industrialists, or in another formulation, of traditionalists over modernizers, to explain the late industrialization of the country under the authoritarian New State. This is nevertheless a rhetoric used by the historical actors themselves, namely self-interested engineers, and it should not be taken at face value. The problem was never a lack of enthusiasm for technology among the leadership of the fascist regime. Engineers were early enthusiastic supporters of the new regime and their views on how the country should be organized were constitutive of the public policies of the New State. The Technical Institute building, an urban landmark of the capital of the New State, confirms engineers' centrality for the regime. The modernist school suggests as well the importance of considering other European experiences, in this case that of Germany, to understand the institutionalization of a radical nationalistic experience in Portugal.

We would like to insist that engineering was not just a tool to implement the policies defined by the ideologues of Salazar's dictatorship. More than that, engineering was an important supplier of the political imagination of the regime. Dams, which we treated more in detail in the text, are a case in point. While irrigation dams promised to solve the demographic problems of the country and promote Portuguese traditional values in carefully planned communities both in Portugal and in its African colonies, models of arch dams for electricity production made plausible visions of Portuguese rivers sustaining large national industries replacing imports. Methods for calculating dams were creatively adapted from American ones developed to reclaim the West.

As in all other examples, Portuguese engineers were invested in reimagining Portugal through methods explicitly learned abroad. After reading the present narrative we hope it will be hard to dispute that engineers were crucial historical actors in imagining Portugal as a nation. It also became clear that this national imagination was only possible by purposefully engaging with international examples learned in Italy, France, England, Germany, and the United States. The historical privileged positioning of engineers in Portuguese society always depended on their exclusive access to international knowledge and skills and how they put these in use for national agendas. Engineers were then the best advocates of their own importance asserting how essential they were to remake Portugal at the image of other European nations and making multiple plans to close the gaps they had first been the first ones to identify.

As briefly mentioned, the new democratic regime had major consequences for the engineering profession in Portugal: not only did it open engineering training to a much larger number of students, women included, as the subjects expanded as well. But we still need more historical research to better understand how engineering practices interacted with the ongoing processes of democratization of Portuguese society. Major dynamics such as the process of integration in the European Union, officially started when the country joined in 1985 the then-called European Economic Community, certainly owe much to engineers. After all, a large part of such integration was materialized through infrastructures projects such as highways, water treatment plants, wind turbines, or fiber optic networks. It is indeed impossible to imagine the country without the cohorts of engineers trained every year at multiple universities.

Nevertheless, their specific role as engineers is harder to identify, since they don't carry with them *the esprit de corps* of their 19th predecessors trained at the Polytechnic School and the Army School, or the ambitions of the Technical Institute engineers of the 1920s and 1930s of being considered among the cultural elite of the country. Like their American and European counterparts, current emphasis on entrepreneurship and innovation rhetoric makes engineers closer to other experts as for example those trained in finance or the biomedical sciences.[15] The blurring of engineering and the sciences is only more obvious for engineers identified with academia, as eloquently demonstrated by Elvira Fortunato, trained as materials engineer at Nova University. Due to her pioneer work on paper electronics, she was the first Portuguese to be awarded an ERC Grant (European Research Council) worth 2.25 million euros, becoming a role model for many Portuguese scientists independently of her identity as scientist or engineer.

---

[15] For an exploration of such changes for the American MIT, see Rosalind Williams, *Retooling: A Historian Confronts Technological Change*, (Cambridge, MA: MIT Press, 2003).

In addition, engineers are still very visible in the Portuguese political system, with no less than four engineers as prime ministers of democratic governments: Alfredo Nobre da Costa, 1978; Maria de Lurdes Pintasilgo, 1979–1980; António Guterres, 1995–2002; José Sócrates, 2005–2011. But with exception of the first, the other three were first and foremost prime ministers who happened to be engineers, which demonstrated mostly the social status attached to the profession and the possibilities of social mobility offered by the formal tittle. They were not politicians acting as engineers, putting engineering at the forefront of their political visions as Fontes Pereira de Melo, Andrade Corvo, Ezequiel de Campos, or Duarte Pacheco. We certainly don't share the elitist politics put forward by these men who offered top-down solutions to engineer the country problems they themselves identified. But, through our long history of engineers, we do see value in reflecting on how engineering practice has been since its origins both a highly political activity and a defining element of Portuguese culture, suggesting the importance of inquiring how engineering might contribute to the current Portuguese experiment with democracy. Unfortunately, the large majority of the curricula of engineering courses in Portugal tends to evade politics and culture as significant subjects for engineers. They forgot the cultural value of *Técnica* for the engineers of the early 20th century or the political implications of industry for those of the 19th century. We hope this book contributes to reassert the political, social, and cultural significance of engineering for any discussion about Portugal.

# Authors' Biographies

## MARIA PAULA DIOGO

**Maria Paula Diogo** is Full Professor of History of Technology and Engineering at the NOVA School of Sciences and Technology and member of CIUHCT - Interuniversity Centre of the History of Science and Technology. She publishes regularly in reputed international journals and publishers. She is a founding member of international research networks, organized numerous national and international conferences and workshops and participates in international H2020 projects. Her most recent co-authored book, *Europeans Globalizing: Mapping, Exploiting, Exchanging* (Palgrave Macmillan 2016) is part of the *Making Europe: Technology and Transformations* book series, which was awarded the Freeman Prize by the European Association for the Study of Science and Technology. She recently co-edited the volume *Gardens and Human Agency in the Anthropocene* (Routledge, 2019). Her research in history of science and technology brings together topics such as globalization, circulation and appropriation of expertise, networks, centers and peripheries and more recently the concept of Anthropocene. In 2020 she was awarded the Leonardo da Vinci Medal, the highest recognition from the Society for the History of Technology.

## TIAGO SARAIVA

**Tiago Saraiva** is an Associate Professor of History at Drexel University, coeditor of the journal *History and Technology*, and author of *Fascist Pigs: Technoscientific Organisms and the History of Fascism* (2016), winner of the 2017 Pfizer Prize awarded by the History of Science Society. He recently coedited *Capital Científica* (Imprensa Ciências Sociais, 2019) and *Nature Remade* (University of Chicago Press, 2021). His work intersects history of science and technology, political history and global history.

Printed in the United States
by Baker & Taylor Publisher Services